BOSTON STUDIES IN THE PHILOSOPHY OF SCIENCE

VOLUME XXXV

BETWEEN EXPERIENCE AND METAPHYSICS

BOSTON STUDIES IN THE PHILOSOPHY OF SCIENCE

EDITED BY ROBERT S. COHEN AND MARX W. WARTOFSKY

VOLUME XXXV

STEFAN AMSTERDAMSKI

BETWEEN EXPERIENCE AND METAPHYSICS

Philosophical Problems of the Evolution of Science

D. REIDEL PUBLISHING COMPANY

DORDRECHT-HOLLAND / BOSTON-U.S.A.

Library of Congress Cataloging in Publication Data

Amsterdamski, Stefan.
    Between experience and metaphysics.

    (Boston studies in the philosophy of science ; v. 35)
    (Synthese library ; 77)
    Translation of Między doświadczeniem a metafizyką.
    Bibliography: p.
    Includes index.
    1. Science—Philosophy—Addresses, essays, lectures.
    2. Knowledge, Theory of—Addresses, essays, lectures.
    I. Title. II. Series.
    Q174.B67   vol. 35 [B67]   501s [501]   75-2184
    ISBN 90-277-0568-2
    ISBN 90-277-0580-1 pbk.

MIĘDZY DOŚWIADCZENIEM A METAFIZYKĄ:
Z filozoficznych zagadnień rozwoju nauki
Originally published by Książka i Wiedza, Warsaw, 1973
Translated from the Polish by P. Michałowski
Edited by R. S. Cohen

Published by D. Reidel Publishing Company,
P.O. Box 17, Dordrecht, Holland

Sold and distributed in the U.S.A., Canada and Mexico
by D. Reidel Publishing Company, Inc.
306 Dartmouth Street, Boston,
Mass. 02116, U.S.A.

Printed in The Netherlands by D. Reidel, Dordrecht

For Olga and Joanna

Reason, holding in one hand its principles to which alone
concordant appearances can be admitted as equivalent to
laws, and in the other hand the experiment which it has
devised in conformity with these principles, must approach
nature in order to be taught by it. It must not, however, do
so in the character of a pupil who listens to everything that
the teacher chooses to say, but of an appointed judge who
compels the witnesses to answer questions which he has himself formulated.

KANT, *Critique of Pure Reason*

If we never sin against reason we cannot achieve anything.

EINSTEIN

EDITORIAL PREFACE

Polish philosophy of science has been the beneficiary of three powerful creative streams of scientific and philosophical thought. First and foremost was the Lwow-Warsaw school of Polish analytical philosophy founded by Twardowski and continued in their several ways by Leśniewski, Łukasiewicz, and Tarski, the great mathematical and logical philosophers, by Kotarbiński, probably the most distinguished teacher, public figure, and culturally influential philosopher of the inter-war and post-war period, and by Ajdukiewicz, the linguistic philosopher who was intellectually sympathetic with the anti-irrationalist (as he would say), logistic and meta-theoretical inquiries of the Vienna Circle. Second was independent and lively Polish Marxism, with its fine development of social research under Krzywicki, a social anthropologist and younger contemporary of Engels, and then after the war the economist Lange, the philosophers Schaff, Kołakowski, Baczko, and many others. Finally there has been a wide range of philosophical, scientific and humanistic scholarship which lends its various qualities to the understanding of both the logic of science and the historical situation of the sciences: we mention only that great and humane physicist Infeld, the phenomenologist with deep epistemological interest Ingarden, the historian of scientific ideas Zawirski, the historian of philosophy and aesthetics Tatarkiewicz, and the mathematical logicians such as Mostowski and Szaniawski. With their sensitivity to the historical materialist sociology of culture, the logical and linguistic clarification of scientific knowledge, the technical progress of modern logic and the physical and biological sciences, our contemporary Polish colleagues who work in the philosophical understanding of science have had an unusual opportunity to bring the different 'moments' of insight into a coherent discourse. And especially the Polish Marxists, for they have been perceptive students, as well as critics, of their analytic colleagues. We should be able to learn from them how to bring together the historical situation of social interests with the logic of concepts and theories, and indeed, going further, how to understand that the nature and role of rationality undergo change as much as the tools, instruments, and purposes of the sciences.

Stefan Amsterdamski has responded to these themes of historical and epistemological inquiry in a series of books and papers over many years. Along with so many others in the East European countries, he has been stimulated by the 'new' historical philosophers of science, by Feyerabend, Toulmin, Lakatos, and above all by Kuhn, and also by the critics of classical logical empiricism, and especially by Popper. His work also reflects that of his own colleagues, among them Eilstein, Pomian, Suszko and Mejbaum. We had his helpful participation in our Boston Colloquia during a long and pleasant part of his Research Fellowship in the United States during 1973–74, and it is our wish to bring Professor Amsterdamski's sober and reflective discussions of recent developments in the philosophy of science to a wider audience through this book.

Professor Stefan Amsterdamski was born in 1929, and completed his higher education at the chemical faculty of the Polytechnical Institute in Lodz and then, for doctoral studies in philosophy, at the University of Warsaw. His dissertation dealt with epistemological problems of the evolution of the notion of chemical element. His habilitation thesis dealt with the objective interpretation of the notion of probability. From 1954 to 1968, Amsterdamski was a member of the faculty of the University of Lodz, for the last four of those years as Chairman. Since 1970, he has been a member of the Institute for the History of Science of the Polish Academy of Science, at Warsaw.

We append a selected list of Amsterdamski's published writings.

Center for the Philosophy and   ROBERT S. COHEN
History of Science,   MARX W. WARTOFSKY
Boston University

SELECTED WORKS BY S. AMSTERDAMSKI

1961   'Rozwój pojęcia pierwiastka chemicznego' [The Evolution of the Notion of
       Chemical Element], Warsaw, pp. 1–216.

1962   'O Obiektywnych interpretacjach pojęcia prawdopodobieństwa' [Objective In-
       terpretations of Empirical Probabilistic Statements], in *Studia Filozoficzne* **2**
       (1962), 139–169, and 3 (1962), 67–101 [French translation in *Studia Filozoficzne –
       Selected Articles* No. 2 – (1964), 3–22].

1964   '*O Obiektywnych interpretacjach pojecia prawdopodobieństwa*', w: *Prawo, koniec-
       zność, prawdopodobieństwo* [*On the Objective Interpretations of the Concept of
       Probability*], Warsaw, pp. 1–130.

1967   'Prawdziwość i prawdopodibieństwo' [Probability and Truth], in *Zeszyty
       Naukowe Uniwersytetu Łódzkiego*, pp. 1–16.

1968   (a) 'Historia nauki i filozofia nauki' [History of Science and Philosophy of
           Science]. Poslowie do *Struktury rewolucji naukowych T. S. Kuhna* [Introduc-
           tion to the Polish translation of T. S. Kuhn's *The Structure of Scientific
           Revolutions*], Warsaw, pp. 197–206.
       (b) 'Wstęp do filozofii' [Introduction to Philosophy], Warsaw, pp. 1–650.

1970   (a) 'Scjentyzm i rewolucja naukowo-techniczna' [Scientism and the Scientific-
           Technological Revolution], *Zagadnienia Naukoznawstwa* **3** (23), pp. 16–33
       (b) 'Spór o problem postępu w historii nauki' [The Problems of Progress in the
           Evolution of Science], in *Kwartalnik Historii Nauki i Techniki* 3, pp.487–506.

1971   (a) 'Nauka a wartości' [Modern Science and Values], in *Zagadnienia Naukoz-
           nawstwa* **1** (25), pp. 58–73 [English translation in *Polish Sociological Bulle-
           tin*, 1973].
       (b) 'Context of Discovery and Context of Justification' [In English], in *Materi-
           ały na Miedzynarodowy Kongres Historii Nauki*, Moscow and Warsaw,
           Editions of the Polish Academy of Science, pp. 95–109.
       (c) 'Nauka jako przedmiot humanistycznej refleksji' [Science as an Object of
           Humane Reflection], in *Studia Socjologiczne* **2** ,pp. 27–54 [English transla-
           tion in *Organon*, 1972].

1973   *Między doświadczeniem a metafizyką* [*Between Experience and Metaphysics*],
       Warsaw, p. 1–252.

TABLE OF CONTENTS

PREFACE TO THE ENGLISH EDITION

Since the time that I finished writing this book two years have passed. During this period many works have appeared, both books and articles, which deal directly with many of the questions raised in this work. In the course of this period I was able, during one year which I spent in the United States, to meet with a number of persons, whose names appear in this book, and to discuss with them a large number of problems of mutual concern. Therefore, when Professor R. S. Cohen and Professor M. W. Wartofsky decided to publish my work in the *Boston Studies in the Philosophy of Science*, I was faced with a difficult problem – in what way to take into account in the English edition the many new works, as well as those to which previously I had no access, and to what extent to include in the text the analysis of problems and polemics voiced in various discussions. The solution which I have chosen had to be, as is usually the case, a compromise between what was possible and what was necessary. The fact that the principal ideas expressed in this book have withstood, as I see it, the brunt of criticism, has led me to remain, basically, with the original text. In the hope that in the future I will return to many of the problems which would extend the scope of the problems discussed in my book, I have decided against any serious expansion of the original text. The corrections, which I have introduced, were aimed primarily at the more precise expression of certain theses, the elimination of errors which have been pointed out to me, and the refutation of certain criticisms, with which I do not agree, but which might also come to the mind of the reader. In addition, I have somewhat expanded the bibliography and added a number of footnotes, either to the new works or to discussions which I had occasion to participate in.

I am most gratefully thankful to both R. S. Cohen and M. W. Wartofsky (Boston University) for their assistance in publishing this book in English as well as for organizing a discussion of a summary of the basic ideas of this book, which I presented to the Boston Colloquium for the Philosophy of Science. I am also most grateful to Professors E. Nagel (Columbia

University), A. Grünbaum and L. Laudan (both of the University of Pittsburgh) as well as Professors D. Da Solla Price and M. Klein (both of Yale University) for enabling me to discuss certain fragments of this book at seminars on the philosophy and history of science at their universities. I am also indebted to all the participants in these discussions for their comments, regardless of whether I agree with them or whether they have found their way into the present edition of this book.

I am in particular most grateful to the late Professor I. Lakatos (London School of Economics), as well as Professors A. Grünbaum (University of Pittsburgh), L. Laudan (University of Pittsburgh), and J. Agassi (Boston University and Tel Aviv University) for meetings, discussions and incisive observations.

Words of thanks are due also to Dr. P. Michalowski for the effort which he put into the translation and for his pleasant cooperation in the preparation of the English edition.

New Haven, July 1974   S. AMSTERDAMSKI

PREFACE TO THE ORIGINAL EDITION

The subject of this book is the philosophical problems of the evolution of science. It is dominated by three problems, and these serve to organize and structure its content:

First of all, the question by what measure is it possible to delimit science from all other products of human intellectual activity, by means of historically invariable methodological criteria which scientific knowledge is to fulfill? In other terms – is it possible for a philosophy of science, which programatically limits itself to the study of logical and methodological problems, to explain what is science, and to construct on this basis an adequate model of its evolution?

Secondly, what is the nature of the relationship between the empirical basis of science (facts), and theory, and whether the changes of this basis (the discovery of new facts and the more precise formulation of already known information) constitute a sufficient foundation for the understanding of the process of the evolution of knowledge?

And finally, the problem of the role which is fulfilled in the evolution of knowledge by philosophical convictions, and especially by epistemological and ontological reflection, and of the impact of this reflection upon the results of cognition, that is upon the content of scientific theories.

These questions also provide the point of departure for the analysis of several more detailed problems which have often been discussed in contemporary philosophy of science.

The central aim of this book however, is an attempt to approach science from such a perspective which would replace the idea of a one-sided impact of facts on theoretical thinking, as well as the reverse notion, according to which facts are shaped by theoretical thinking, with a view of science as a game between the mind and experience; and by that token to treat science as a fragment of a larger whole – the intellectual culture of a given period. The reader must judge whether this aim has been reached.

Fragments of this book, as well as the problems raised in it, were the subject of a seminar which I directed in 1970–71 and 1971–72 at the Department of the History of Science and Technology of the Polish Academy of Science. Several chapters were the subject of lectures at the Warsaw section of the Polish Philosophical Society and at methodological seminars at Warsaw University and the University of Poznań. I wish to thank warmly all those who took part in these discussions, which helped me to clarify my own opinions on many points.

I am particularly grateful to the first readers of the manuscript of this book, Prof. Klemens Szaniawski and Dr. Krzysztof Pomian. By their comments and suggestions they helped me at many points to improve the text and to eliminate a number of errors. Particularly, I want to warmly thank K. Pomian for many private, friendly conversations on problems raised in this book. They amount to a debt which is not easy to repay.

Warsaw, July 1972   S. Amsterdamski

CHAPTER I

RADICAL EMPIRICISM AND THE
ANOMALIES IN THE KNOWLEDGE OF SCIENCE

I

One of the essential features of man which makes him different from all other living beings is his capacity for self-reflection. This capacity enables him to reflect on what he does and how he should act, as well as what he thinks and how he should think. These two spheres of consciousness – knowledge and self-knowledge – are intertwined with each other to such an extent that it is practically impossible to separate them. This circumstance has tremendous import for any study of the products of human culture, and, therefore, also for the study of science. In addition to the fact that the results of any such study usually become, or at least try to become, a part of the very universe under study, this universe is by itself a complicated fabric of knowledge and self-knowledge, of science and knowledge of science. It follows that one of the goals of such a study consists in unraveling the mutual relationships and dependencies between them. When inquiring into the nature of science, we have to deal with a specific system of views about the world and, at the same time, with a certain set of opinions about that very system. If we should choose to call such theoretical reflection the *philosophy of science*, we must be aware that its object consists of scientific as well as of philosophical opinions; namely, of those philosophical opinions which are implicitly contained in scientific knowledge, and of those which concern the way of practicing scientific investigations, i.e., methodological and epistemological opinions. Such a state of affairs does not in any way simplify research. However, it appears impossible to ignore it, for that leads, as we shall attempt to demonstrate, to basic simplifications and misconceptions in the understanding of science.

II

As a result of its own dynamic development, as well as its growing practical role in all aspects of life, science has become today the subject of extensive theoretical investigations. These studies can be seen as oriented in two fundamental directions. On the one hand, they deal with organization of scientific research and with science policy, which have become important factors in the economic and social planning of all nations. The goal of these investigations is to ensure the proper tempo and organizational ramifications of research and to secure the practical application of scientific achievements. On the other hand, science is an object of investigation of the traditional humanities, such as history, sociology, psychology and now the methodology of science.

It is usually taken for granted that these disciplines aim at the understanding of science as a fundamental element of human culture. Thus, history of science analyses the development of scientific knowledge in the past, the problems which science has had to confront, the ways in which it has solved them. All these historical problems are investigated, if possible, within a consideration of the vast social context in which science has evolved. The sociology of science studies the evolution of science as a social institution, the styles of scientific thinking, the reception of scientific ideas and their social determination. Finally, logic and the methodology of science (often called the *philosophy of science*), as a branch of epistemology, study the structure of scientific theories, their development, the rules of concept and theory formation, the criteria of accepting and refuting the claims of science, and the relations between theory and empirical data.

The borders between all these disciplines are, in practice, often vague. Such specialization is a natural phenomenon in the development of any system of knowledge, not only of the knowledge of science. Usually it leads to a deeper understanding of particular aspects of phenomenon under study. It is well known, however, that such specialization also has negative effects. They are the more dangerous, the more direct is the mutual relationship between the particular aspects of the phenomenon under study; that is, when it is impossible to reach an understanding of any of them in isolation without studying their mutual relations. Today, more than ever, we are aware of the fact that "specialization is not a virtue, but a necessary evil, and only has a value if it leads in the long run to the integration of knowledge."[1]

Such is the situation we face today in theoretical reflection upon science, especially in the study of its development. It is more and more widely recognized that a deeper understanding of the process of the growth of knowledge is impossible without overcoming the negative effects of excessive specialization in studies on science. This objective, we believe, is one of the tasks of *philosophy of science*.

On the grounds of the last statement it should be clear that the term *philosophy of science* will be understood here in a broader sense than is customary in contemporary literature, which usually limits this discipline to the problems of logic and methodology.[2] We believe that its task is to investigate science as a particular domain of human intellectual activity, as a specific component of human culture, and that it must go beyond logic and methodology. The understanding of mutual relations between the methodological standards of scientific activity and the socio-historical conditions in which science is practiced should be considered as one of the most important tasks of the philosophy of science. We believe, in fact, that none of the traditional disciplines studying science (methodology, history, sociology) can, on its own, draw a coherent picture of the growth of knowledge, though all these disciplines have a common object of study. However, without attempting a synthesis of the results obtained by these disciplines, it is impossible to properly understand the specific problems they deal with.

The choice of the term *philosophy of science* for denoting this kind of reflection about science is, of course, an arbitrary terminological decision. Though we have chosen this term in view of certain general convictions we hold concerning the tasks and the object of philosophy, as long as the choice remains a purely arbitrary one, there is, we believe, no need to justify it. It is not the choice of the term that requires justification, but the proposed orientation of our inquiry. It is hoped that this book will fulfill just this requirement.

III

Anyone who has even a slight interest in the history of science must have found himself faced by many bewildering problems. How was it possible, for instance, that the most remarkable minds of their times held opinions that could today be disproved by any high school student, often on the basis of empirical facts known even then? It is just this last circumstance which appears to be most striking, for it is one thing to accept theories which later may be proved imprecise or simply false, but are not contradicted by known facts, and quite a different matter to maintain opinions regardless of known facts which clearly contradict them. If the first instance seems quite natural to us, as we believe that it is the discovery of new facts which usually compel us to revise old theories, then the second instance appears to be perplexing; in fact, it seems to contradict the very nature of scientific knowledge.

No fact, however, is strange by itself. If it is unusual and strikes us as an anomaly, that is because it does not fit into our scheme of how 'it should be'. Since we usually derive such schemes from philosophical and popular literature on science, it seems appropriate to analyze more closely the basic assumptions of a view of science which leads us to see such facts as anomalies.

Some of these assumptions can be exemplified, at least in a cursory manner, by comparing common views on science and on art.[3]

The bewilderment that we experience when confronted with certain facts from the history of science is parallel to a certain extent to the emotions we feel when for the first time we are faced with the art or customs of a culture that is different from our own. We find it haid to believe that men could behave so 'irrationally' or that they could hold such 'strange' aesthetic values. But an expert, as opposed to a layman, knows perfectly well that it makes little sense to evaluate the products of another culture by the same criteria that he would apply to the products of his own culture. When studying primitive art, or for that matter the art of any other era, he will attempt to understand the principles and criteria which determined the creators at their epoch and not to judge their products by today's standards. He is aware, in any case, that these criteria need by no means be equivalent.

It is not the same situation, however, in the case of understanding science. Here not only the layman but also the expert often holds the view that in the history of science the criteria of rationality have remained unchanged, i.e., that they have a supra-historical character and are immanently connected with science in the course of history. This opinion often serves as the basis for the delimitation of science from other products of human intellectual effort: on the grounds of this opinion, the products of cognitive activity that do not conform to these criteria are considered to be unscientific.

Although the art historian would consider it senseless to ask whether the paintings of the Aztecs were more valuable than those of the European Renaissance, the philosopher or historian of science sees nothing wrong with comparing the theories of Ptolemy and Copernicus, Newton and Einstein, or Darwin and Mendel. He is convinced that these theories had to provide a coherent explanation of the same domains of phenomena (the movement of celestial bodies, mechanics, the mechanism of heredity), and that in attempting to formulate explanations they utilized the same criteria for evaluating the results of their work.

Although we usually refrain from formulating such judgments as that the *Venus de Milo* is more beautiful than Rodin's *Thinker* (and if we encounter such opinions we consider them simply as expressions of individual artistic taste which we are not obliged to accept), we nevertheless agree that modern physics is more perfect or 'truthful' than medieval physics. We add only that "we are wiser because we are standing on the shoulders of giants". When admiring the paintings of the Impressionists we do not by the same token deny any credit to the works of Rembrandt. But when we accept the theory of relativity, we *eo ipso* deny the truth of certain statements and assumptions of pre-relativistic physics. In the course of the history of science certain theories replace others, sending them to the archives of science or to history textbooks. In art, on the other hand, the appearance of new masterpieces does not compel us to reevaluate the old ones: these are not to be replaced by the new. Thus, we believe that it is impossible to speak of progress in the same sense in art as in science. We would agree that Newton as a scientist would have to recognize Einstein's theory of relativity, but that Leonardo, as an artist, would not necessarily have to like the paintings of Kandinsky or Klee, nor to paint in their style.

Although we are prone to evaluate works of art in a relativistic manner, taking into account the aesthetic criteria and historical circumstances which are particular to a given period, we tend to evaluate the achievements of scientists (just their *achievements*, not their merits) on the grounds of some supra-historical criteria. More precisely, we tend to evaluate scientific results on the grounds of those criteria which are accepted by contemporary science and which we take (for better or for worse – we will come back to this problem) as supra-historical. This means that we see science as a process which is continuously aiming at the same objective, namely truth, and what is more, we believe that the very concept of truth was always understood in the same manner. Furthermore, we see science as a process in which its claims are always accepted or refuted on the grounds of the same criteria (conformity with experimental results, first of all), as a process whose every stage constitutes a logical consequence of the previous one, and in turn is itself the logical antecedent of the following stage. Science is then perceived as cumulative process in which new truths are continuously being added and old mistakes successively eliminated. The goal of science – the acquisition of true knowledge – remains constant; so do its methods and criteria, namely confrontation of theoretical knowledge with empirical data, which are believed to be independent of any theory. The mechanism of the growth of knowledge is seen as a process in which new theories replace old ones as soon as these are in conflict with empirical data, but have to take over from them all those statements which were not empirically refuted. What changes in this process is only the content of our knowledge, the scope of our information.

Regardless of the degree in which this view can be considered as valid (and admittedly it is hard to deny that it strikes at certain basic characteristics of science as, for example, those that differentiate it from art) it remains clear that is assumes a particular philosophical conception of the evolution of knowledge, which we shall call *radical empiricism*.[4] In accord with our statement above, we must take into account that this conception, which belongs to the domain of knowledge *of* science, does not remain passive in respect to science itself, but interferes with scientific activity and cannot be separated from it, especially in modern times. At the same time it cannot be denied that in the light of this conception, a number of historical facts appear as anomalies, as exceptions to the pattern which this view proposes. If this is so, then it follows that what we take for the pattern needs to be corrected at certain points. There seems to be no reason to assume that the theory of knowledge and of its evolution should not correspond to the facts provided by the history of science.

IV

How can we explain the fact that for centuries scientists believed that an arrow, shot from a bow, moved as a result of the air pressure caused by the released bowstring. We know that this opinion resulted from the theory according to which a body can move only as long as a force is exerted upon it. But it was well known at the time (it was impossible not to know) that an arrow could fly up-wind, against the direction of the 'moving air'. How was it then, that for centuries this fact was not taken as a sufficient condition for refuting the accepted theory of motion?

Anyone who assumes that science consists in unbiased observation of phenomena and the inference of logical conclusions from them, cannot but be surprised that such a theory could have been formulated at all. Anyone who assumes that scientific theories are accepted *solely* on the basis of their conformity with empirical data cannot but wonder how such a theory could have been formulated, as well as how it could have been accepted (with only small revisions) for almost two millennia.

Later, however, when due to the work of Descartes and Galileo the idea of inertia was introduced, it became apparent that the revision of the previous theory did not require either the use of new, more precise instruments, or new observations. What was necessary was to look at the problem from a completely different perspective. It was enough to assume that inertial motion as well as the state of rest is a natural state of a body and the whole problem could be seen in a new light. But this seemingly small change meant nothing less than the rejection of the whole of Aristotle's cosmology and a basic reconstruction of the whole world-view based on this cosmology.

Anyone who assumes that new theories are formulated and old ones abandoned exclusively under the pressure of new empirical data cannot grasp the basic quality of that intellectual revolution in which the rise of modern physics played such an important part. He does not see, in effect, that the long-lasting acceptance of contradictions between current theory and known empirical data, as well as the subsequent abandonment of this theory without any new empirical facts, had their roots in a common factor, namely in the extra-scientific beliefs of the leading authorities of that era.

The example given above, which is meant to demonstrate the need for revision of the radical empiricist theory of knowledge, according to which theoretical changes are caused only by empirical data, can easily be questioned. It may be claimed, for instance, as some philosophers and historians have, that given the speculative character of ancient and medieval thought, it is impossible to speak of science prior to the 16th–17th centuries when the authority of experiment was finally established. If it were so, the disputed thesis of radical empiricism could not be refuted on the grounds of the history of ancient or medieval physics which then, strictly speaking, was not scientific and can be treated only as the pre-history of genuine science.

Although there can be no doubt that the 16th and 17th centuries were a period of profound intellectual revolution which in many ways deter-mined the shape of modern science and its subsequent evolution, we cannot accept this argument as a defense of radical empiricism.

First of all, let us remark that the argument according to which the history of medieval physics cannot contradict the thesis of radical em-piricism, because this physics was not scientific, is based on the trans-formation of a descriptive statement into a definition which tells us how the term '*science*' has to be used, what may and what may not be denoted by it. No matter whether the disputed thesis does or does not apply to modern science, it is nevertheless impossible to defend it by means of a definition of science which depends on this very thesis.

But let us disregard the formal aspect of the problem: the use of a wrong argument does not yet discredit the thesis it is supposed to defend. We do not accept the thesis of radical empiricism because, on the one hand, we claim that it does not fully apply to modern science, and be-cause, on the other, it is an exaggeration to state that ancient and medieval science did not accept the authority of experiment and obser-vation.

We might add that it is hard to understand how particular authors reconcile their views on the speculative, anti-empirical nature of ancient and medieval thought and the thesis that modern science was born out of the process of overcoming Aristotelian physics and philosophy. Rather one should be surprised that so many precise and important observations were made without use of advanced instruments and that such ingenuity was utilized in relating these observations and theoretical constructs. How can we accept the idea that medieval science did not take into account data obtained by observation when we know that the Ptolemaic theory was constantly being improved and revised and that in order to make it more precise and conforming with observations, new epicycles, deferents and eccentrics were introduced into it? And if in astronomy the results of observation were taken into account[5], why should they not have been utilized in other domains of knowledge? Is it not more reasonable to believe that this statement is an oversimplification of the facts? Should not we rather investigate what was the impact of the accepted theory upon the appraisal of empirical data as important or unimportant, and how the experience itself was understood?

It is worth noticing that the idea that the Renaissance upheaval in science consisted simply in the acceptance of the authority of experiment is a source of many a historical legend. Thus we hear of the elderly Galileo tugging heavy loads up the Tower of Pisa and dropping them off in order empirically to prove the falseness of the Aristotelian theory that the time of fall depends on the mass of the falling body. But, in fact, we know that Galileo never performed that experiment. It was one of his opponents who used this experiment in order to disprove Galileo's theory, while Galileo discarded the results of this experiment claiming that the measurements were not precise enough.[6] And little reflection is needed to see that the undertaking of such experiments lay rather within the traditions of Aristotelian empiricism than within the sphere of Galileo's empirico-mathematical method. The problem of the empiricism of modern science founded by Galileo, and of the speculative character of medieval science represented by the Aristotelians, is undoubtedly much more complicated than the current textbooks would have us believe. The aforementioned legend tells us more about popular views of science than it does about the nature of Galileo's method, which was to have such a profound impact on the subsequent development of science and which in reality had little in common with empiricism as it is so understood. In fact, this legend is a product of seeing science through the lens of radical empiricism.

Even if for the sake of argument we would agree to date the beginning of science to the intellectual upheaval of the 16th and 17th centuries, we could still easily demonstrate that such cases, as are exemplified by the theory of the flying arrow, happened also in later times. And, to cite one example, how may we explain that during the 16th and 17th centuries biological heredity was explained in terms of the theory of preformation? According to this theory the organisms of all future generations preexist in miniature, one within the other, while the act of insemination only triggers the process of growth. This strange theory was meant to explain the similarity of organisms of subsequent generations. At the same time it raised a furious discussion as to whether the preformed organism exists

already in the sperm or in the ovum. It is not astonishing that to the arguments of opponents who stressed that it was impossible to discover the
supposed preformed organism by the best microscopes then available,
Malebranche answered: "The mind should not stop at what the eye sees,
for the vision of the mind is far more penetrating than the vision of the
eye."[7] Still, we know that some biologists claimed to have seen these
miniature organisms by microscope. It is hard to believe that they could
have seen them if they had very much not wanted to. As Fontana rightly
commented: "Anyone can look into the microscope, but only a few can
truly judge what they have seen."[8]

How was it possible, however, to defend the theory of preformation in
the face of such evidence as cases of cross-breeding which could have
hardly remained unobserved? And what was one to do with an animal that
resembled neither of its progenitors and could preexist neither in the
sperm nor in the egg? "How could pre-existence and preformation be
reconciled with the unforeseen circumstances of copulation?"[9] Why, we
must ask, did the question not arise until the end of the 17th century, from
where in a series of preformed organisms does, what we today would
call a hybrid come?

Are we then to judge that the theory of preformation was simply
unscientific? That would in no way change the fact that it belonged to the
science of that time. And if it should be treated as unscientific, then what
of the 19th century theory of the ether which endowed this medium with
contradictory features?

Does the argument of Malebranche not bring to mind the arguments
advanced in the 19th century in defense of the atomic theory, which was
being attacked by radical empiricists, although on different philosophical
grounds?

If we treated as scientific only those theories which are compatible with
all accepted empirical data, we would have to judge that the heliocentric
theory of Copernicus was also unscientific. In effect, it was in conflict
with all the accepted "facts" of contemporary physics, which Copernicus
did not question. There is no evidence that he considered it as inadequate.
On the contrary, he attempted to reconcile his theory with accepted
physical conceptions. It is enough to read in *De Revolutionibus*[10] the
answers given by Copernicus to those who (on the grounds of science, not
of theology) questioned the theory of the movement of the earth in order

to realize that all 'rational' (from the point of view of contemporary knowledge) and 'empirical' (from the point of view of the understanding of facts imposed by this knowledge) arguments were squarely against him. He had to defend himself on enemy territory, that is on the grounds of Aristotelian physics, and on that ground there was no possible defense. During the course of almost a hundred and fifty years, that is until the time of Newton, the Copernican theory faced a multitude of physical anomalies which it could overcome only with great pain, and not always. The theory could be defended on rational grounds only after the Newtonian breakthrough. Until that time, the followers of Copernicus could have been considered as cranks by established science.

The following passage is taken from a book which, as a result of its atheistic tendencies, found itself on the papal *Index Librorum Prohibitorum*. The author, Jean Bodin, states:

No one in his senses, or imbued with the slightest knowledge of physics, will ever think that the earth, heavy and unwieldy from its own weight and mass, staggers up and down around its own center and that of the sun; for at the slightest jar of the earth, we would see cities and fortresses, towns and mountains thrown down. A certain courtier Aulicus, when some astrologer in court was upholding Copernicus' idea before Duke Albert of Prussia, turning to the servant, who was pouring the Falernian, said: "Take care that the flagon is not spilled". For if the earth were to be moved, neither an arrow shot straight up, nor a stone dropped from the top of a tower would fall perpendicularly, but either ahead or behind...[11]

Such arguments were commonly advanced against Copernicus, who had no rational – either in the contemporary, or the modern sense – defense. He could not defend himself on the grounds of Aristotelian physics, and from the point of view of modern physics his arguments can in no way be considered as rational.[12] He defended his theory without any new empirical data which could contradict the old theory. Even worse, he defended it against those 'facts' which were perfectly well explained in terms of Ptolemean system. (The prediction of stellar parallax, which was incompatible with the geocentric system, was not empirically confirmed until much later.) In fact, if Copernicus has been guided in his work by even the most liberal criteria of the radical empiricist theory of knowledge, he would have never arrived at his theory. But, of course, we know that he was right and that he is rightly considered to be one of the greatest minds in human history. As Einstein used to say: "If we never sin against reason we cannot achieve anything."[13] This very reason, in effect, even though it appeals to what are usually treated as 'unquestionable facts', sees them in the light of contemporary theories.

It was not only the Copernican theory that for years had to deal with a mass of anomalies. The same thing happened with Newton's theory, with quantum mechanics, and in general with any new theory which required a radical reconstruction of large areas of acquired knowledge. This is the fate of any new theory which attempts to introduce a new order into the domain of phenomena which it describes and which were ordered in a different way by its predecessors. Since every such theory must take into account experiments and observations previously carried out, it must therefore reinterpret them, make them fit into the new theoretical framework. Sometimes this reinterpretation is extremely complex and hardly discernible to anyone but the specialist, especially in cases when the meaning of those basic concepts which became a part of everyday language undergo a radical change. Thus after the rise of the theory of relativity, the meaning of such notions as 'mass', 'velocity' or 'simultaneity' had changed, and this fact should be taken in account when we investigate its relation to Newtonian mechanics.

We see, therefore, that in some cases, when empirical statements are taken over by a new theory, they achieve a new meaning. The process of such reinterpretation takes time. Even assuming that the new theory is better than the old one, empirical facts still seen through the view of the old theory continue to appear as anomalies. This is especially the case when these concepts have filtered through into public consciousness. It was known for years that the observed orbit of Mercury was incompatible with Newton's theory and yet no one rejected it because of this fact. It was believed that just as in many cases of previous anomalies the theory will be able to overcome this one, too. But what are the reasons for such an expectation? The elimination of apparent anomalies and the transformation of data that seem to contradict the theory into confirming evidence is one of the most essential tasks of scientists who accepted the new viewpoint, who trusted in it as fruitful. The more general was the old theory, the more numerous were the implications it had for other fields of research, and the larger the amount of anomalies faced by the new one. Whether it will be able to overcome them depends not only on the scope of the known empirical facts[14] but also on many other factors as well, such as, for instance, the contemporary state of mathematical knowledge, the theoretical situation in other disciplines, the state of the technical apparatus available or the accepted epistemological and ontological beliefs, etc. To quote F. Jacob, there is "... a domain which thought strives to explore, where it seeks to establish order and attempts to construct a world of abstract relationships in harmony not only with observations and techniques but also with current practices, values and interpretations."[15]

In all the examples we have cited, from the case of the flying arrow to the problem of the orbit of Mercury, we have been dealing with one and the same fact: that a theory is maintained in spite of contradicting empirical facts. It appears that in each case the pattern is the same; because of the trust invested in the theory, it was assumed that empirical data was only apparently in contradiction with the theory and that in time they would be proven compatible with the new order that it constituted. This trust is not only based on the past successes of the theory in dealing with empirical data, but also on a number of other factors which do have impact upon the appraisal of theoretical constructions, but which are not taken into consideration by radical empiricism.

It is not, therefore, surprising that anyone, who in a simplified way sees the relationship between facts and theory, who believes that a scientific theory is accepted when it is compatible with all known empirical data and that it is refuted as soon as contradictory data appears, must of necessity find in the history of science many anomalies and deviations from what he would consider to be the normal and obligatory pattern of the growth of knowledge. For anyone who sees the evolution of science in that light, there remains no alternative but to consider these anomalies as aberrations in the normal operative scheme or at the best as exceptions to the norm. The number of these 'exceptions' is, however, much too large to be treated as 'exceptions which confirm the rule'.

V

It is an indisputable fact that the coming into existence and shaping of modern science evolved under the credo of observation and experiment and the dismissal of all the metaphysical speculations which had been treated as characteristic of scientific thought in the preceding centuries. Empiricism was indeed an article of faith for many modern scientists and a methodological program advocated by many philosophers. At the same time, the successes of science, especially of the physical sciences from the time of Galileo to Newton, Laplace and Maxwell, appeared to confirm the belief that it is the constant clash between theory and experience that constitutes the primary impetus for the development of knowledge. In this clash, which often leads to the formulation of new theories and abandoning of old ones, the dynamic revolutionary factor seems to lie in the sphere of experience, theory remaining a conservative element. The roots of this attitude lay, firstly, in the belief that facts, and only empirical facts, determine the content of scientific theory and, secondly, that the facts to which we refer in the process of testing a theory compose an unquestionable basis for knowledge: they are unequivocal and free from any theoretical bias. Claude Bernard exemplified this concept of scientific method when he wrote that: "Now to find truth, men of science need only to stand face to face with nature and... question her with the help of more and more perfect means of investigation."[16]

In dozens of scientific and philosophical books we may read that thanks to Bacon, Galileo, Newton and many others, scientists came to believe that only meticulous observation, free from any theoretical bias, can deliver us from sterile hypothesising and empty disputes and assure the growth of knowledge. Let us see then to what extent the history of sciences confirms that point of view, according to which solid scientific theories arise only on the ground of empirical data.

It is well known that the Copernican theory was not the first heliocentric system in the history of astronomy. The concept of such a system was advanced in the third century B.C. by Aristarchus, and already a century earlier Heraclides of Pontus defended a similar theory. The idea that the earth rotated around the sun was an element of the astronomical and cosmological theories of the Platonists and Pythagoreans, and Copernicus himself quoted them in the introduction to *De Revolutionibus*. However, until the Renaissance it was the geocentric system, as elaborated in detail by Ptolemy, and which was compatible with the cosmology of Aristotle, which was dominant. We know that Copernicus needed neither telescope nor any new observation in order to reject this theory. It was simply necessary to look at the known facts in a new way. We must, therefore, seek to find which factors led to the situation that the heliocentric theory was not formulated at the time of Ptolemy, that it had to remain dormant for centuries to be rediscovered by Copernicus. The empirical basis for both theories was, after all, the same.

It is worth remembering that the empirical confirmation of the Copernican theory, which at the same time disproved the Ptolemaic system, was only provided many years after the death of Copernicus. Copernicus himself could offer no decisive fact which would confirm his own theory. "No fundamental astronomical discovery, no new sort of astronomical observation, persuaded Copernicus of ancient astronomy's inadequacy or of the necessity for change."[17] The prognoses of celestial phenomena provided by either theory proved to be essentially the same. In any case, in the sixteenth century, Copernicus could not have claimed that his theory could provide more precise predictions. The reform of the calendar, which was meant to serve as a motive for the search for a new theory, could only be carried out years after his death, as a result of much intensive work by his successors. The only argument that Copernicus could advance in favor of his theory was that it was mathematically simpler. This was precisely the defense of the theory which was chosen by Osiander in the famous preface to *De Revolutionibus*[18] and which Cardinal Bellarmine offered to Galileo. Galileo, however, had at his disposal empirical facts which spoke on behalf of the new theory and, what is more important, was better prepared to reconcile that theory with contemporary physics than Copernicus had been

It is, therefore, difficult to say that the Copernican theory was less 'speculative' or more 'empirical' than its predecessor. It is impossible to explain by means of pure facts (in this case astronomical) why the heliocentric system had not been elaborated and accepted before Copernicus, nor why he himself chose to develop it. In order to understand these phenomena it is necessary to reach out beyond the 'naked facts' and even beyond astronomy. Without going into details,[19] let us simply note that one of the principal factors which motivated Copernicus and Kepler, and to a certain extent also Galileo, to reform astronomy were their neoplatonic philosophical convictions.

Neoplatonism completes the conceptual stage setting for the Copernican Revolution.... For an astronomical revolution it is a puzzling stage, because it is set with so few astronomical properties. Their absence, however, is just what makes the setting important. Innovations in a science need not be responses to novelties within that science at all.... Any possible understanding of the Revolution's timing and of the factors that called it forth must, therefore, be sought principally outside of astronomy, within the larger intellectual milieu inhabited by astronomy's practitioners.[20]

VI

Isaac Newton is usually considered as an ardent partisan and representative of the empirical method, which provided science with hitherto unencountered successes. But does the statement '*hypotheses non fingo*' with which he explained the lack of an explanation for the essence of gravitation, really account, as some would have it, for the way by which he formulated the principles of the new physics? Is it really correct to treat this theory as an inductive generalization of observed empirical facts discovered without any prior assumptions or hypotheses, even without any assumptions and hypotheses which transgressed the boundaries of physics? Or was this only a program, an ideal which even he himself could not realize, as seems to be confirmed by the famous appendices to the successive editions of his *Optics* and the *Scholia* appended to the *Principia*?

By introducing into physics such strange notions as 'momentary action at a distance', distinguishing between absolute and relative time and space, endowing all of nature with a capacity such as gravitation (which he himself perceived as immaterial and which was to reconcile the concept of inertial, rectilinear motion with the observed path of celestial bodies) was Newton really generalizing on the basis of performed experiments? Or is it rather the case that not without connection with his philosophical and theological views he constructed a conceptual apparatus which was to interpret all physical experiments? Whatever the case, we know that his rules, which were 'derived from facts', which opposed metaphysical and speculative hypotheses, were not so convincing for contemporary scholars that they should accept them without any discussion. A century had to pass before these rules were accepted by scholars on the continent.[21] By claiming, then, that he was not presenting hypotheses, was Newton truly reporting his method? Or was it that by presenting the mathematical relationships which he had arrived at as facts and not as hypotheses he was in reality defending himself against the arguments of his opponents, who questioned his theoretical synthesis mainly on philosophical grounds? And were his famous polemics with the Cartesians and with Leibniz only arguments between an empiricist physicist and metaphysicians, or were these in fact disputes between natural philosophers, who on the grounds of physics disputed God and the universe?

We know, thus, that although Newton was able to provide a physical criterion which would differentiate between absolute and relative circular motion (namely the existence of centrifugal forces), he was unable to provide such a criterion for the simplest form of motion – rectilinear motion. And although he could not demonstrate how to differentiate experimentally between absolute and relative rectilinear motion, and knew the Galilean principle of relativity, he nevertheless maintained his concepts of absolute space and time and the resulting theory of absolute motion. And what inclined him to maintain these concepts? A. Teske writes:

Did Newton conceive of it as a postulate making it possible to describe motion, or as a mere assumption taken for granted for methodological reasons? The truth was that Newton thought of his view as certain, but his certainty was of metaphysical, not of empirical origin.[22]

The notebooks of David Gregory, a pupil and friend of Newton's, contain a passage which clearly informs us that Newton considered his metaphysical and theological conceptions as safeguards of the logical consistency of his physical system.

What the space that is empty of body is filled with, the plain truth is that he [Newton] believes God to be omnipresent in the literal sense; and that as we are sensible of objects when their images are brought home within the brain, so God must be sensible of everything, being intimately present with everything: for he [Newton] supposes that as God is present in space where there is no body, He is present in space when a body is also present.[23]

Thus it seems we may rightly state that if the 'metaphysician' Descartes deduced the law of the conservation of motion from the concept of the invariability of God, then the 'empiricist' Newton in turn deduced his ideas of absolute time and space from the notion of the ubiquity of God. And this was the meaning of the famous statement, that time and space are God's *sensorium*.

As A. Koyré has convincingly demonstrated,[24] the dispute between Newton and both the Cartesians and Leibniz concerned in equal measure both physics (the concepts of time, space, gravitation, atomism) and metaphysics – that is the place and the nature of the influence of God in the functioning of the mechanism he had himself created. As may be seen quite readily, this aspect of the debate was very much in the open, as Leibniz, for example, wrote:

Sir Isaac Newton, and his followers, have also a very odd opinion concerning the work of God. According to their doctrine, God Almighty wants to wind up his watch from time to time; otherwise it would cease to move. He had not, it seems, sufficient foresight to make it a perpetual motion. Nay, the machine of God's making is so imperfect, according to these gentlemen, that he is obliged to clean it now and then by an extraordinary concourse, and even to mend it, as a clockmaker mends his work; who must consequently be so much the more unskillful a workman, as he is oftener obliged to mend his work and to set it right. According to my opinion, the same force and vigour remains always in the world and only passes from one part of matter to another, agreeably to the laws of nature, and the beautiful pre-established order.[25]

Thus, Leibniz, in defending the law of the conservation of motion also appeals to a specific theological conception.

Newton's *porte-parole*, Clarke, replied thus:

Sir Isaac Newton doth not say, that space is the organ which God makes use of to perceive things by; nor that he has need of any medium at all, whereby to perceive things: but on the contrary, that he, being omnipresent, perceives all things by his immediate presence to them, in all space wherever they are, without the intervention or assistance of any organ or medium whatsoever....

The notion of the world's being a great machine, going on without the interposition of God, as a clock continues to go without the assistance of a clockmaker, is the notion of materialism and fate, and tends (under pretence of making God a *supramundane intelligence*) to exclude providence and God's government in reality out of the world. And by the same reason that a philosopher can represent all things going on from the beginning of the creation, without any government or interposition of providence, a sceptic will easily argue still further backwards, and suppose that things have from eternity gone on (as they now do) without any true creation or original author at all, but only what such arguers call all-wise and eternal nature.[26]

And in reply to Leibniz's second letter, Clarke wrote: "... so far as metaphysical consequences follow demonstratively from mathematical principles, so far the mathematical principles may (if it be thought fit) be called metaphysical principles."[27] Thus wrote the *porte-parole* of the author of the famous aphorism 'hypotheses non fingo' on the dispute over the nature of physical space, time and on motion and the law of its conservation.

When a century later Laplace carried the new cosmology to perfection, and could answer Napoleon on the role which God fulfilled in his *Systems of the World* with the famous statement "*Sire je n'ai pas eu besoin de cette hypothèse*", it was in one sense but a Pyrrhic victory of Newtonianism over Cartesianism and Leibniz. For as Koyré has so rightly noted,

Every progress of Newtonian science brought new proofs for Leibniz's contention: the moving force of the universe, its *vis viva*, did not decrease; the world clock needed neither rewinding, nor mending.

The Divine Artifex had, therefore, less and less to do in the world. He did not even need to conserve it, as the world, more and more, became able to dispense with this service.

Thus the mighty, energetic God of Newton who actually 'ran' the universe according to His free will and decision, became, in quick succession, a conservative power, an *intelligentia supra-mundana*, a *Dieu fainéant*.[28]

We could at this point add that not only the development of Newtonian physics, but also its 'downfall' after Einstein was a confirmation of Leibniz's correct concept of the relative nature of time and space and, therefore, of the principal point of argument between Newton and Clarke.[29] The work initiated by Leibniz of bringing to light the metaphysical assumptions of the *Principia Mathematica Philosophiae Naturalis* was completed by the author of *Zur Elektrodynamik der bewegten Körper*. Although at the time of the triumphs of Newtonian physics the declaration of its founder against hypothetical statements were taken as an adequate account of his method by empirically minded scientists, in the context of the disputes which he led with his opponents at the beginning of the eighteenth century, they had quite a different sense. For they were directed against a specific metaphysics and particular physical hypotheses, but at the same time they defended a different metaphysics. This was quite apparent both to the Cartesians and to Leibniz, and also to Voltaire, who had occasion to seek information at the source (he had personally met with Clarke and Cotes) and openly acknowledged this fact in the first part of his *Elements*.[30]

To quote Teske once more:

... any vagueness [about the metaphysical background of Newton's conception of space and time] is dismissed by a careful reading of Voltaire's book. It could have performed this role even before Gregory's notebook was discovered, if only it had been given enough attention. Newton's ideas have constituted the foundation of mechanics for more than two hundred years. That Newton consciously linked them with metaphysics is not only interesting, but also important for any consideration of the point of departure of the theory of relativity.[31]

Thus the physical inconsistencies of Newton's system, veiled by metaphysics, remained unnoticed before the formulation of the theory of relativity until the system was triumphant, and metaphysical opponents, who could not provide any alternative theory, had to remain silent.

In time scholars became accustomed to these inconsistencies, learned not to notice them, and the concept of absolute time and space was rationalized through Kant's philosophy. Physics and metaphysics were, within this system, as in any physical system, so intimately linked with each other that it took the revision of its basic ideas, after the rise of the theory of relativity, to prompt the inquiry into the source of Newton's concepts of time and space. And it does not seem to be the case that the earlier discovery of Gregory's notes might have changed this situation to any extent. For the answer to this question could be found in almost any text written at the beginning of the eighteenth century, when Newton's mechanics had not yet finally dominated the field – both in Leibniz's polemics with Clarke as well as in Voltaire's *Elements*, and last, but not least, in the very *Principia* and *Optics* of Newton himself. It could not be found, however, in the presentations of Newtonian physics dating from the period when it was triumphant, when it was presented as an ideal product of the application of the empirical method. This radical interpretation was so widespread, that Engels referred to Newton as an 'inductive ass'. Thus, in this concrete historical example, empiricist philosophy stood in the way of the philosophical criticism of science and hindered the discernment of its assumptions which lay outside the sphere of science. The content of Newtonian mechanics not only stepped outside the boundaries of empirical data and, therefore, could not be deduced from them, but also was incompatible with observational data which were known at the time when it was formulated – such as the impossibility of distinguishing between absolute and relative motion.

VII

Thus it is not the case that modern science as forged in the intellectual revolution of the 16th and 17th centuries, and whose crowning glory was Newtonian physics, constituted a practical embodiment of the empirical method, at least in the interpretation which forces us to treat scientific theory as the extract from experience. This is in equal measure as untrue as the statement that earlier science was purely speculative and did not take experiment into account.

It is true, however, that beginning with the seventeenth century, such an interpretation of empiricism was propagated and considered the only scientific one, and that scholars, including such great ones as Newton, attempted to present their discoveries in such a way as though they had, in fact, reached their results in the manner prescribed by the empiricist philosophers. In fact, one may truly say that "this [classical empiricism] is characterized by a kind of schizophrenia. What is propagated and declared to be the basis of all science is a radical empiricism. What is *done* is something different."[32] Presenting oneself as a partisan of the empirical method belonged to the *savoir-vivre* of scholars who were anti-metaphysically minded, belonged in a manner of speaking to the fashion of the times. By this token the philosophers who promoted this fashion could, with satisfaction, search out confirmation of their views in the work of scientists. Without doubt this is not the only example of a situation when each creative achievement of the human mind has been interpreted as 'the product of the only true philosophy', and when scholars, sometimes just in order to be let alone, and sometimes out of conviction, paid declarative homage to it. In any case, neither statements of scientists nor reasonings of philosophers provide sufficient confirmation of the thesis, that in posing experimental questions about nature, and reading the answers by means of measuring instruments, scientists were free from hypotheses and presuppositions and that in declaring the glory of radical empiricism they were not subject to any philosophy or metaphysics. We would indeed be quite naive if we took their words in this matter. And especially in light of what Bertrand Russell has noted, that "whatever presents itself as empiricism is sure of widespread acceptance, not on its merits, but because empiricism is the fashion."[33] This 'fashion', of course, had its origins. In the minds of the partisans of empiricism, experiment was to guarantee the autonomy of science, was to free it from the tight bonds of philosophical and theological dogma and to free it from any outside authority.

It required the analyses of a number of historians of science, such as Duhem and Koyré, and above all the rise of the theory of relativity, of quantum mechanics and the great epistemological discussion on the origins of physics and modern science in general, to bring to light the fact that the radical empiricist conception of science and its evolution were in need of scrupulous reappraisal.

The selected historical examples we have cited above seem to indicate that this analysis should primarily direct itself to the following problems:

First of all, to what extent is it justified to seek to delimit science from all other products of human intellectual activity by means of methodological criteria, which scientific knowledge must satisfy?

Secondly, what is the true relationship between the empirical basis of science (i.e., facts), and theory, and what changes of this basis can provide a satisfactory foundation for the understanding of the process of the evolution of scientific cognition?

And finally, to what measure are the opinions justified that the evolution of knowledge may be explained as a process of systematic accumulation of truth and elimination of mistakes through the confrontation of theories with facts, and that in reality successive scientific theories must be linked by the relationship of correspondence, as is claimed by the methodology of radical empiricism?

These questions, to which the following investigations are dedicated, undoubtedly constitute the central problems of the philosophy of science. Therefore, in accordance with what we have already said, our analysis has to proceed on two levels: those of science and those of the knowledge of science.

NOTES

[1] E. Schrödinger, *Science et Humanisme*, Paris 1954, p. 21.
[2] This opinion will be analyzed in Chapter II.
[3] See Chapters II and III.
[4] For a more detailed analysis see Chapter IV.
[5] Cf. O. Neugebauer, *The Exact Sciences in Antiquity*, Boston 1957.
[6] H. Butterfield, *The Origins of Modern Science, 1500–1800*, New York 1957, ch. 4.
[7] Malebranche: *Recherche de la Vérité*, Paris 1700, t. 1. p. 48.
[8] Cf. E. M. Wermel: *Istoria uczenia o kletkie*, Moscow 1970, p. 25.
[9] F. Jacob: *The Logic of Life, A History of Heredity*, Pantheon Books, New York 1974, p. 68.
[10] N. Copernicus: *De Revolutionibus...* Book I, Chapters VIII, IX.
[11] T. S. Kuhn: *The Copernican Revolution*, Harvard University Press, Cambridge 1966, p. 190.
[12] Cf. A. Koyré, 'Le problème physique du copernicanisme', in *Études Galiléennes*, Paris 1966, pp. 166–171.
[13] Cf. B. G. Kuznetzov, *Albert Einstein*, Moscow 1962, p. 110. [English translation, Moscow, 1965, p. 104; original text in *Lettres à Maurice Solovine*, Paris 1956, p. 128. – Ed.]
[14] Sometimes a great quantity of facts may be an obstacle for a theoretical generalization. It seems dubious whether Mendeleev could formulate his law of periodicity if more chemical elements were known in his times. The amount of anomalies would have been much greater if he knew all the rare elements. The same seems true in

respect to Mendel. Perhaps he would not have formulated his law if he had experimented not only with peas, whose hereditary features are transmitted independently one of another, because the statistical relations would have been more complicated.

[15] F. Jacob, *op. cit.*, p. 11.

[16] C. Bernard, *An Introduction to the Study of Experimental Medicine*, New York 1957, p. 221.

[17] T. S. Kuhn, *op. cit.*, p. 131.

[18] Osiander's foreword is usually interpreted either as an expression of his conventionalistic attitude, or as a quibble for preventing condemnation by the Church. No matter what was his motivation, it is by no means easy to answer the question how, in 1543, the Copernicus theory could be defended otherwise.

[19] Cf. T. S. Kuhn, *op. cit.*; H. Butterfield, *op. cit.*; A. Koyré, *From the Closed World to the Infinite Universe*, Baltimore 1957; *Études Galiléennes*, Paris 1966; 'De l'influence des conceptions philosophiques sur l'évolution des théories scientifiques', in *Études d'histoire de la pensée philosophique*, Paris 1961, pp. 231–247.

[20] T. S. Kuhn, *op. cit.*, p. 131.

[21] Cf. Pierre Burnet: *L'introduction des théories de Newton en France au XVIII siècle*, Vol. I, Paris 1931.

[22] A. Teske: 'Voltaire's *Elements of Newtonian Physics*', in *The History of Physics and the Philosophy of Science*, Warsaw 1972, p. 78 [in English].

[23] *Ibid.*, p. 80 [quoted from Teske's source; Philipp Frank, *Einstein*, New York 1947, p. 35. – Ed.]

[24] A. Koyré, *From the Closed World to the Infinite Universe*, Baltimore 1957, Chapters IX, X, XI.

[25] 'Mr. Leibniz's First Paper', in *The Leibniz-Clarke Correspondence*, ed. by H. G. Alexander, Manchester 1956, pp. 11–12.

[26] 'Dr. Clarke's First Reply', *ibid.*, pp. 12–13, and 14.

[27] 'Dr. Clarke's Second Reply', *ibid.*, p. 20.

[28] A. Koyré, *From the Closed World...*, p. 276.

[29] Cf. Leibniz Third Paper, *ibid.*

[30] Cf. Voltaire, *Eléments de la philosophie de Newton, mis à la portée de tout le monde*, Chapters I and II.

[31] A. Teske, *op. cit.*, p. 80.

[32] P. K. Feyerabend, 'Problems of Empiricism', in *Beyond the Edge of Certainty*, Prentice Hall Inc., Englewood Cliffs, N.J. 1965, p. 154.

[33] *Ibid.*, p. 145. [Quoted from *The Philosophy of Bertrand Russell*]

CHAPTER II

TROUBLES WITH THE PROBLEM OF DEMARCATION

I

Any theoretical reflection on science must, implicitly or explicitly, base itself on a more or less articulated definition of science. Thus, for instance, it is impossible to study the history of science without accepting in advance some opinion concerning the scope of the phenomenon, the evolution of which one attempts to investigate. When a historian of science begins his investigation with Antiquity or with the Renaissance, when he includes within, or excludes from, the universe under study such domains of knowledge as magic, astrology, alchemy, mathematics or philosophy, he expresses, by the same token, his opinion of the question: 'what is science', even if he does not mention this problem explicitly. The same is true, of course, in the case of the psychology or the sociology of science.

It is not hard to demonstrate that these opinions are taken, consciously or unconsciously, from some philosophy of science. Thus, the manner in which philosophy explains what is science has impact on other disciplines studying science, even if the philosopher in question does not hold any pretentions as to the acceptance of his concept of science by studies of science which are oriented otherwise. Therefore, regardless of the intentions and goals determining the way in which particular philosophies construct the concept of science and specify the criteria of scientific method, they are, nevertheless, in some manner 'responsible' for the vision of science functioning in sociological or historical studies and for its common-sense understanding.

Trying to answer the questions 'what is science?' and 'what are the genuine criteria of scientific method?', contemporary philosophy of science has been largely preoccupied with the so-called *problem of demarcation*.

In this chapter, we shall analyze the attempts to solve this problem, as well as some issues directly related to them.

II

As Popper stated:

> The problem of finding a criterion which would enable us to distinguish between the empirical sciences on the one hand, and mathematics and logic as well as 'metaphysical' systems on the other, I call the *problem of demarcation*. This problem was known to Hume, who attempted to solve it. With Kant it became the central problem of the theory of knowledge. If, following Kant, we call the problem of induction 'Hume's problem', we might call the problem of demarcation 'Kant's problem'. Of these two problems – the source of nearly all the other problems of the theory of knowledge – the problem of demarcation is, I think, the more fundamental.[1]

It follows that the solution of the problem of demarcation has to provide a criterion by which it would be possible to state what does and what does not belong to the realm of science. This criterion specifies the *potential* boundaries of science, because it provides the conditions which must be satisfied by any system of scientific claims, of those which have already been formulated as well as of those which will be formulated in the future.

It goes without saying that the boundaries of science so delimited would depend on the chosen criterion. Or, to put it otherwise, the choice of the criterion depends on what we want to exclude from the realm of science. Thus, Popper's claim, quoted above, makes it clear that the criterion which he is looking for has to exclude logic and mathematics from the realm of science. This is a consequence of the fact that he is looking for a criterion delimiting empirical sciences, and of his opinion that logic and mathematics do not belong to them.

The solution of the problem which we have just presented has been indeed one of the essential goals of philosophy of science during the last century. Mach, Poincaré and Duhem attempted to solve it and so, later, did Wittgenstein (in his *Tractatus*) as well as the members of the Vienna Circle – Schlick, Carnap and Reichenbach. The problem became the central concern of Popper's methodological reflections on science. Although he rejected the solutions proposed by his predecessors, he nevertheless inherited from them the very problem, as well as certain philosophical convictions which caused him to take up the question.

The results of these attempts are well known and have been widely discussed in philosophical literature. We shall, therefore, limit ourselves to a short recapitulation of the proposed solutions without discussing them in detail.

We must begin by mentioning the *criterion of verification*, first formulated by Wittgenstein in his *Tractatus*, and then elaborated by Schlick and
Carnap.[2] According to this conception, the meaning of any statement
depends upon the possibility of its verification, i.e., upon the possibility
of proving it to be true. So, we have two kinds of judgments: those which
can be proven true and which are called *meaningful*, and those which
cannot; not because they are false, but because they are outside of any
method of control. These are called meaningless or metaphysical. The
truth-value of meaningful statements can be demonstrated either empirically, or by means of the analysis of meanings of their components. In
the first case they are empirically verifiable, while in the second, according
to Kant's terminology, analytically verifiable. The statements of empirical
sciences belong to the first category, the statements of formal sciences
(logic and mathematics), to the second.

This conception, as its authors soon realized themselves, could not
withstand criticism and had to be abandoned. In addition to many other
difficulties, it implied that scientific laws are not meaningful statements:
being universal statements, they cannot be conclusively proved as true,
but only confirmed to some degree by empirical evidence. However, even
a strong confirmation does not exclude the possibility of future empirical
falsification.

As a consequence, Carnap[3] and Reichenbach[4] replaced verifiability by
a more liberal criterion of confirmation (or testability). Retaining the
basic differentiation of judgments into metaphysical, empirical and
analytical, they assumed that the scientific status of statements (their
meaningfulness) depends upon the possibility of confirmation by empirical evidence. Because of various difficulties concerning the empirical
confirmation of theoretical statements[5] (i.e., of statements including
expressions which denote unobservable entities) the conception of
confirmability as a criterion of meaning underwent several modifications resulting in further liberalisation of the proposed criterion of
demarcation.[6]

Both of these conceptions were undoubtedly based on the assumptions
of radical empiricism, especially on the assumption that the status of
scientific statements is determined by the degree of their inductive confirmation, and that experience constitutes an unquestionable basis for
accepting or refuting the claims of science. Thus, statements giving account

of the results of observations were treated as foundations of the whole structure of scientific knowledge.

It was Popper, rejecting inductivism and the conception of a purely empirical basis of science, who advanced another solution of the problem of demarcation.[7] According to him, the necessary condition for scientific status consists in the possibility of falsification. Statements which do not satisfy this condition, i.e., those which can never be falsified by any experiment (whatever would be its result) were qualified as metaphysical, *but not as meaningless*. As we intend later to analyze the Popperian conception in a more detailed fashion,[8] we will mention here only the principal arguments advanced against it. First of all, the criterion of falsifiability excludes from the realm of science all universal existential hypotheses, as well as all probabilistic statements. In the same way as universal statements cannot be conclusively verified, so existential hypotheses cannot be conclusively falsified. As far as statistical statements are concerned, they can neither be conclusively verified, nor conclusively falsified by experiment. Whatever would be the statistical distribution obtained in an experiment, it may be always treated as a more or less probable fluctuation. As we will try to demonstrate later, falsificationism is also incompatible with the fact that the empirical falsification of the claims of science is neither a sufficient nor a necessary condition for their elimination.

All three solutions of the problem of demarcation may be criticized because they exclude logic and mathematics from the realm of science. This criticism is advanced by those authors who do not accept the analytic-synthetic dichotomy as absolute.[9]

It should be perhaps noticed that some authors, for example, H. Mehlberg,[10] try to combine both criteria (of confirmability and falsifiability), and assume that the scientific status of a statement depends upon the possibility of their confirmation *or* falsification.

III

Despite the differences among the aforementioned solutions of the problem of demarcation, they have some common features that are worthy of notice. First and foremost, all of them are inspired by the idea of a demarcation of science which would guarantee that scientific status will be attributed only to those statements which are susceptible to the verdict of experiment. This idea has its roots in the vast tradition of modern empiricism, and, while it surely grasps some important features of scientific knowledge, it is nevertheless a difficult one to realize. This is so, first of all, because the criterion which we are looking for should not exclude from the realm of science, knowledge which, according to the opinion of specialists, belongs to it; secondly, because it should not exclude from science too much of what passed for science in the past. However, it turned out that the proposed criteria were too rigorous (sometimes too liberal) in this respect, which means that they could not accomplish their methodological functions. The normative conception of science, which inspired the search for a solution of the problem of demarcation, could not but fall into conflict with the postulate demanding a non-arbitrary delimitation of scientific knowledge.

Generally speaking, these difficulties may have two different causes. First, they might result from the fact that the criterion of demarcation has not yet been satisfactorily formulated and, therefore, it does not accomplish its tasks; secondly, it may be that the problem, as it is formulated, is not solvable at all because it springs from some false premises.

We believe that we are dealing with the second case. The search for the line of demarcation is based on the assumption that it is possible to specify criteria of scientific status which have a supra-historical character, which do not change with time. If this were not the case, no criterion could delimit science *tout court*, but only science of a particular historical epoch. Moreover, the search for the criterion of demarcation has its source in the belief that the empirical nature of scientific knowledge consists in the fact that only those statements which are susceptible to the verdict of experiment belong to it. If it were so, the delimitation of these statements would be equivalent with drawing a demarcation line between science and other products of human intellectual activity, and, what is more important, that the drawing of this line might provide a basis for understanding the pattern of the process of growth of knowledge. We shall attempt to demonstrate that both of these assumptions are wrong.

At the same time, however, we should realize that attempts to answer the question 'what is science?' by delimiting it from other products of human intellectual activity are determined by a manner of evaluating science as a social phenomenon, by opinions concerning the functions it fulfills or should fulfill in the social life, and by the need for an exact demarcation of the scope of its responsibility and competence. This need springs mainly from proliferation of doctrines which pretend to be scientific, but which do not satisfy even the most elementary methodological criteria accepted by scientific communities. The danger of such mystifications grows stronger with the rise of the social reputation of science and with its specialization, due to which the layman is not able to judge for himself the reliability of doctrines which label themselves as science and which various media propose that he accept. It seems, in fact, that never in the past has there been such a large gulf between common and specialized knowledge as exists today, and that never has the reputation of science among non-specialists been based to such an extent on confidence and faith. This undoubtedly is a situation which provides ample room for the abuse of this reputation. The need to separate science from doctrines which parasitically live off its reputation, is one of the prime motives inspiring the search for a clear-cut criterion of demarcation, especially by men who prize science highly, who see it as a supreme expression of human rationality. The empirical character of scientific knowledge may seem to be the feature which appears to be the most promising for the establishing of the line of demarcation. Therefore, even if these attempts to solve the problem of demarcation have not led to a satisfactory solution, even if the concept of science based on these premises has turned out to be defective, it is nevertheless hard to deny that they have helped to keep a watchful eye on the various self-styled representatives of science and have served a useful role as foundations for scientific deontology. They have "awakened intellectuals to their own responsibilities and in my opinion have been of practical aid in counteracting attempts to blur the boundaries between the position of the scientist and the obligations of the believer."[11]

Let us try to analyze the difficulties common to all of these solutions.

IV

At first glance it may appear that the solution of the problem of demarcation should not present any serious difficulties. It would suffice, it might seem, to state which common features are to be found in scientific statements as opposed to non-scientific ones. A moment of reflection, however, should suffice to let us understand that such a road leads nowhere. For in order to indicate what differentiates scientific statements from any other, we must first delimit them, and to do so we must, of course, use some criterion of demarcation, which is precisely what we are looking for. Hence, it appears that whatever criterion of demarcation we would formulate, whatever feature of statements we would choose as a symptom of their scientific status, our criterion cannot be formulated as a descriptive statement. In this sense it is impossible at all to answer the question, 'what is science?'.

Whenever it appears to us that we may answer this question, we are, in fact, wrong. We simply do not perceive that we are either introducing into language a normative definition of the type 'by the term *science* we mean this or that' or 'we shall call *science* this or that', or we report on a particular linguistic usage of the term and state how it was used in a particular period of time.

In the first case our answer has the character of a linguistic convention. Such conventions are in common use in all the domains of knowledge whenever a clear definition is needed. In the second instance, although we do formulate a descriptive statement, it is not, however, a statement about science, but about the linguistic usage of the term *science*. It is not hard to see that this usage can be fluid and determined by a large number of factors, and no great historical knowledge is needed in order to realize that, in fact, the term was utilized in different ways in the past. Nevertheless, in answering the question as to how the term *science* was used is by no means the same as solving the problem of demarcation. We are simply providing the answer to a different question.

If it is so, then the criterion of demarcation regardless how we formulate it, must be of normative character, and, by the same token, its acceptance or refutation is always a matter of convention. In such a situation, any rational discussion can pertain only to the usefulness of the norm, and this is possible only under the condition that there is an agreement as to the ends which the accepted norm is to serve. The choice of these ends is always a matter of decision which cannot be rationally questioned;[12] to justify a norm is to refer to some particular value, while the justification of the acceptance of this value must refer to another value which is considered *autothelic*, i.e., as one that does not require any further justification. At this point, of course, any rational discussion must end. For us, however, it is important to notice that, as in any choice of norms of behavior so in the choice of methodological norms, we are dealing with some system of values. It is at this point – contrary to the illusions of radical empiricists concerning the possibility of delimiting knowledge from self-knowledge – that methodology loses its ideologically neutral character and becomes linked to some particular system of values. And, to the degree in which science cannot be delimited from its methodology, it connects science with the acceptance of some systems of values.

Hence, whenever we state that a certain theory is, or is not scientific, whenever we say that certain domains of knowledge (philosophy, mathematics, logic, or even metaphysics) do, or do not belong to science, whenever we express an opinion on when science was born, we do this on the ground of some conventionally accepted definition of science. That this definition might at one time be almost universally accepted, that it may seem to be the only correct and 'rational' one, does not change its conventional character, at most it can only serve to mask it.

If we then agree that demarcation criteria are of a normative character, then, in evaluating them, we must take into account what are the goals they are to serve within the frame of a given methodology.

In criticizing the Carnapian conception of the demarcation of science (as a system of empirically confirmable statements) from metaphysics (consisting of empirically untestable, meaningless ones), Ingarden formulated the following question: To what kind of statements does the proposed criterion of demarcation (or of meaningfulness) belong? On what grounds are we forced to accept it? Ingarden demonstrated[13] that the proposed criterion does not satisfy the requirements that it imposes on meaningful statements, and is, therefore, itself meaningless, metaphysical.

It is important to mention that Carnap at that time still held the belief which Popper later characterized as naturalistic, that

every linguistic expression purporting to be an assertion is either meaningful or meaningless; not by convention, or as a result of rules that have been laid down by convention, but as a matter of *actual fact, or due to its nature, just as a plant is, or is not, green in fact, by nature and not by conventional rules.*[14]

Later Carnap rejected this view and replaced it with the opinion that the meaningfulness of a statement depends on the rules of the language to which it belongs, and that "the given expression is a meaningful sentence in certain artificial languages if, and only if, it complies with the rules of formation for well-formed formulas or sentences in that language".[15] With this new position Ingarden's charge could be answered, that the criterion of demarcation is a normative rule, conventionally imposed on the language of science and that only those sentences which comply with this rule can be considered as scientific. This is equal to saying that the criterion of demarcation is a conventional one, one that is not based on any 'natural' features of a sentence, i.e., that any demarcation criterion must be accepted as a norm, and as such may be accepted only as a convention. With such an understanding of the criterion of demarcation, it is impossible to demand of the sentence by which we express the criterion that it satisfies the requirements which it imposes on scientific statements.

It has been remarked above that Popper always rejected the naturalistic point of view and considered his criterion of demarcation and other methodological rules as normative.

This view, according to which methodology is an empirical science in its turn – a study of the actual behavior of scientists or of the actual procedure of 'science' – may be described as *'naturalistic'*…. I reject [this point of view]. It is uncritical. Its upholders fail to notice that whenever they believe themselves to have discovered a fact, they have only proposed a convention. Hence the convention is liable to turn into a dogma. This criticism of the naturalistic view applies not only to its criterion of meaning, but also to its idea of science and, consequently, to its idea of empirical method.[16]

The emphasis on this point of view is of importance here, as we shall be coming back to it repeatedly in the course of our study. It is, however, important to note that it is by no means universally accepted. Thus, for example, K. Ajdukiewicz distinguished between an a-pragmatic and a pragmatic methodology.[17] The former was considered to be a descriptive discipline whose most developed branch is the theory of deductive systems. As for pragmatic methodology, which is undoubtedly the sphere within which Popper is working, it consists, according to Ajdukiewicz, (1) in the delimitation of the procedures utilized in the practice of science and in their analysis, which lead to definitions giving account of what these procedures are; (2) in the general description of procedures applied in particular sciences; (3) in the discerning of tasks, which consciously or unconsciously are undertaken by scientists in various disciplines, and "the resulting codification of sound scientific procedures".[18] The tasks described under (1) and (2) undoubtedly lead to descriptive judgments, not to normative ones. It seems that according to Ajdukiewicz, the tasks specified in (3) lead also to descriptive statements, because the formulation of 'sound procedures' is to be based on the analysis of goals actually aimed at by scientists.

Popper's point of view is decidedly different from the one outlined above. He also analyses how scientists really proceed, but the goal at which he aims is not an analysis of the conscious or unconscious goals of scientific enterprise, but the formulation of methodological norms concerning how science should be done and how rationally to proceed in view of goals that science has to aim at. "It is only from the consequences of my definition of empirical science, and from the methodological decisions which depend upon this definition, that the scientist will be able to see how far it conforms to his intuitive idea of the goal of his endeavours."[19] He stresses that the methodological rules he proposes are consequences of his normative criterion of demarcation. The same point of view was defended by Imre Lakatos, who explicitly said that "philosophy of science provides normative methodologies".[20]

One problem, however, is clear. On the grounds of a descriptive methodology of science, no criterion of demarcation can be formulated. When answering the question 'what is science?', this methodology can tell us that science is what the scientists are doing and that scientists are men who are considered as scientists in their epoch. In other words, when we want to define science on the grounds of a descriptive methodology, we must rather refer to some historically determined concept of science than to supra-historical criteria, which are supposed to delimit it from other products or domains of human intellectual activity.

V

It is not always acknowledged that because of the normative character of criteria of demarcation we are faced by the following problem. When, on the grounds of the conventionally accepted methodological norm, we draw a demarcation line between scientific and unscientific statements, we are, by the same token, establishing another border line too, namely that between science and its methodological rules. It turns out that, because of their normative character, the criterion of demarcation, as well as other methodological rules (rules of accepting and refuting the claims of science, of interpreting experimental results, of constructing concepts and theories, etc.) do not belong to the set of statements denoted as scientific. In consequence, the universe of scientific knowledge is not only delimited from the ocean of unscientific statements which do not satisfy the criterion of demarcation, but is delimited also from whatever is 'above', from the whole of its methodological superstructure, without which science never existed and cannot exist. It becomes delimited from meta-science and from epistemological opinions which obviously interfere with the process of doing scientific research. The delimitation of pure cognitive procedures or of their results from opinions and beliefs concerning their character, methods according to which they should be performed, and their aims, seems to be as impossible as the delimitation of knowledge from self-knowledge. (And this is a point of prime importance not only in the philosophy of science, but in philosophy in general.) Even if the differentiation between a language and a meta-language, between theories and meta-theories, between science and meta-science, is indispensable from the logical point of view, nevertheless a demarcation line which excludes from the realm of scientific knowledge 'the means of its construction and appraisal', i.e., its whole methodological superstructure, seems artificial. Such a demarcation can but provide a distorted picture of science. No matter what were the unquestionable differences between the logical status of empirical and methodological statements, both kinds of them are constitutive components of science and factors determining its development. I would say that meta-science is a part of science in the same way as meta-language is a part of language or self-knowledge is a part of knowledge. Therefore, I think that the principal questions, which must be solved in order to achieve an understanding of science and of the pattern of growth of scientific knowledge, do not concern the demarcation of scientific statements from other products of human intellectual activity, but the mutual links and impacts among the different spheres of theoretical thought.

The situation we have to deal with is almost the same as that when logical and mathematical statements are excluded from science because of their analytical character, because they may be judged as true (or false) on the ground of linguistic rules independent of the results of any experiment. In both cases the demarcation aims at the same goal; science has to embrace only these statements whose refutation or approval depends uniquely upon experimental results. As we said, all the cited solutions of the problem of demarcation were inspired by this goal. I do not intend to discuss here the problem whether the empirical-analytical dichotomy may be sustained in the absolute sense; the problem is controversial.[21] However, it is one thing to state that the logical status of mathematical and logical statements is essentially different from the logical status of empirical claims, and quite another thing to draw the demarcation line according to this difference. In any case, if we delimit science according to this conception, the problem of the pattern of the evolution of knowledge cannot be satisfactorily solved. Not every real difference is worthy of being taken as a basis for demarcation. *The conception according to which science is a system of statements of the same logical status* (from the point of view of the differentiation into empirical, analytical and normative ones) *appears to be wrong.* This opinion will be argued in a more detailed manner when we will discuss the problem of growth of knowledge.

When criticizing the radical empiricists' opinion concerning the empirical analytical dichotomy, and concerning the understanding of science, Quine states:

... total science is like a field of force whose boundary conditions are experience. A conflict with experience at the periphery occasions readjustments in the interior of the field. Truth values have to be redistributed over some of our statements. Reevaluation of some statements entails reevaluation of others, because of their logical interconnections – the logical laws being in turn simply certain further statements of the system, certain further elements of the field. Having reevaluated one statement we must reevaluate some others, which may be statements logically connected with the first or may be the statements of logical connections themselves. But the total field is so underdetermined by its boundary conditions, experience, that there is much latitude of choice as to what statements to reevaluate in the light of any simple contrary experience. No particular experiences are linked with any particular statements in the interior of the field, except indirectly through considerations of equilibrium affecting the field as a whole.[22]

It seems that what Quine says about the statements of logic applies also, *mutatis mutandis*, to methodological rules. Although, on the grounds of the proposed criteria of demarcation, both kinds of statements should be excluded from the universe of scientific knowledge, nevertheless, they are actually components of the system of knowledge called *science*. Both of them may, as the history of science testifies, change as the whole changes. For both of them it is true that "there is much latitude of choice as to what statements to reevaluate". We would only remark that the laws of

logic, which state the connections between statements, and the methodological rules are such components of the system which are reevaluated
only *in extremis*, i.e., when there is no other possibility to reestablish the
equilibrium of the whole field which has been disturbed by the last conflict
with experimental results. This is why they change very seldom, why they
are the most stable components of the system. It is why it may seem that
the system as a whole evolves and works permanently on the grounds of
the same rules, and that it is possible to provide a definition of science or
to solve the problem of demarcation by specifying the methodological
rules which are always to be satisfied by any scientific statement. In order
to realize how deeply logical and methodological rules are involved in
changes of the field, it would be enough to remind the reader of the
discussions concerning the necessity of a new logic as a result of the
paradoxes of quantum physics, concerning probabilistic inferences, or of
the polemics concerning operationalism, instrumentalism and realism with
all their impact upon the discussion of the concept of truth.[23]

We should also notice that when we are saying that "the total field is so
undetermined by the boundary condition that there is much latitude of
choice as to what statements to reevaluate in the light of any simple
contrary experience", we are immediately faced by the question "on what
depends the decision which statements are to be actually reevaluated?".
The answer of this question is primordial for the understanding of the
process of growth of knowledge. Whatever this answer may be, it is
obvious that the question itself cannot even be asked within the frame of
radical empiricism.

Thus, in attempting to answer the question 'what is science?', we are
faced by the following alternative: either we define science through
reference to some normative rules or criteria of demarcation, rationality
and so on, and we assume that they are historically permanent and do not
belong themselves to the universe which they delimit and which evolves
due to their application; or we treat them as 'elements of the changing
field'. In this case, however, they cannot accomplish the function of
delimiting scientific knowledge because they change as the field changes.
In other words, either we treat the criteria of scientific status as permanent,
and then they may be used as means for demarcation of science, or we
treat them as 'elements of the changing field', and then their changes as
well as the evolution of the field must be explained. This explanation has

to specify their dependence upon the 'boundary conditions of the field'. But what I mean by 'boundary conditions' is not only the actually acquired empirical data, but also the socio-historical conditions in which the system evolves being a fragment of the contemporary intellectual culture.

I do not intend to say that we have to choose one of these attitudes for all our investigations of science. In studying the scientific knowledge of some historical epoch we may roughly assume that the methodological criteria on which it is based do not change. But this assumption is misleading when we want to discover the general pattern of the growth of knowledge. In this case, each of the mentioned attitudes leads to a different conception of science and of its evolution and, by the same token, to a different program of studying it.

There is a twofold relationship between epistemology and the special sciences. The former, according to its constructive claims, is fundamental to all the special sciences since it supplies the basic justifications for the types of knowledge and the conceptions of truth and correctness which these others rely upon in their concrete methods of procedure, and affects their findings. This, however, does not alter the fact that every theory of knowledge is itself influenced by the form which science takes at the time and from which alone it can obtain its conception of the nature of knowledge. In principle, no doubt, it claims to be the basis of all science but in fact it is determined by the condition of science at any given time. The problem is thus made the more difficult by the fact that the very principles, in the light of which knowledge is to be criticized, are themselves found to be socially and historically conditioned. Hence their application appears to be limited to given historical periods and the particular types of knowledge then prevalent.[24]

If we assume that the rules of scientific method are permanent, the process of the growth of knowledge is seen only as a series of changes in its *content*, as a succession of theories which are always the results of the same research program. It is clear enough that if the initial assumption is wrong, then as a consequence the reconstruction of the evolution of knowledge cannot be true.

If the methodological criteria are treated as 'elements of the changing field', the growth of knowledge is perceived not only as a series of changes in the content of knowledge, but also of changes of the methodological rules and of research programs. In the first case, the task of historians of science is reduced to discovering who introduced the changes in the content of knowledge, and when they were introduced, as well as to describing and explaining "the congeries of error, myth, and superstition that have inhibited the more rapid accumulation of the constituents of the modern science text."[25] The whole process of the growth of knowledge is seen as a process of accumulation of new information, as the replacement of the old theories by new, more precise ones. In the second case the historians of science "rather than seeking the permanent contributions of an older science to our present vantage... attempt to display the historical integrity of that science in its own time,"[26] that is, they try to grasp the growth of knowledge from the point of view of their contemporary methods, research programs and the goals for which they aimed.

VI

We may ask, however, if such an historical approach to the investigation of science as has been proposed by Kuhn and others[27] does not mean, in reality, a resignation from an attempt to answer the question 'what is science?', what differentiates it from all other systems of human beliefs.

On the one hand, it appears that if we were to say simply that science consists of true knowledge, we would be forced to exclude from consideration almost all past knowledge. All knowledge which we today see as out-of-date would have to be considered unscientific. However, it is self-evident that such a solution is merely a verbal one.

The more carefully they study, say, Aristotelian dynamics, phlogistic chemistry, or caloric thermodynamics, the more certain they feel that those once current views of nature were, as a whole, neither less scientific nor more the product of human idiosyncrasy than those of today. If these out-of-date beliefs are to be called myths, then myths can be produced by the same sorts of methods and held for the same sorts of reasons that now lead to scientific knowledge. If, on the other hand, they are to be called science, then science has included bodies of belief quite incompatible with the ones we hold today, Given these alternatives, the historian must choose the latter. Out-of-date theories are not, in principle, unscientific because they have been discarded.[28]

What is more, it is very probable that future generations will, by analogy, be forced to consider much of our theories in the same manner, for even we ourselves do not consider them to be the last word in the search for truth. In taking this argument to its most extreme consequences, we should be forced to state that science never existed, does not exist, and never shall.

For the same reasons it is impossible, as I have indicated above, to define science on the basis of methodological criteria whose requirements are to be fulfilled by all scientific statements. To do so would be to assume

that these criteria are permanent and that the delimitation of those statements that are susceptible to the verdict of experiment is equivalent with
demarcating the sphere of science from other products or domains of
human intellectual activity.

On the other hand, however, if the rules of scientific method are not
permanent, if science in different historical periods differs not only in the
content of accepted theories, but also in the methods of research, then
in what manner can we speak of science as a historically stable phenomenon? By what right do we use the name *science* for such dissimilar systems
of belief as, for example, Greek cosmology and the theory of relativity,
the theory of preformation and contemporary molecular biology, the
atomistic theory of Democritus and the contemporary theory of the
structure of matter, in other words, for all that at different periods was
considered as science?

It is not hard to state that science continually evolves, that its internal
structure, methods of research, the relationships to technology, degree of
mathematisation, and social role all change from period to period. It is,
however, a much more difficult task to find some stable element, one that
would allow us to call by one name such disparate phenomena as the
knowledge of Antiquity, the Middle Ages, as well as of our own times.
What then allows the follower of the historical approach to embrace all
these phenomena under one *genus proximus*?

The scholar who is dedicated to the historical approach will state, that
rather than search for permanent, normative definitions of science, we
should instead search for "what in different periods men had in mind when
they called a specific field of human activity science, on what basis they
judged activity as scientific, what changes did the notion of science undergo
from antiquity to modern times, and what changes did the moving of the
scientific method undergo."[29] In proposing such a way of investigation,
he must hope that in answering these questions he will come up with
something that will allow him to state that he is in reality studying the
same phenomenon in its various historical manifestations that will allow
him to speak of the self-identity of the subject under study.

The problem which we face is this: is it possible in any way, and if so in
what way, to speak of the self-identity of the subject which we have come
to name *science*? This is an analogous problem to the one which was
discussed by the Athenian Sophists in the example of Theseus' ship.

Let us note first of all, that we can speak of *absolute identity* only in the case of phenomena which do not change in any respect. At the moment that we seek the self-identity of a phenomenon that undergoes some change, we can only speak of some *relative self-identity*. To do this we must find some *principle of individualisation*, that is a criterion which will allow us to state that regardless of any change, we are still dealing with the same object.

The principle of individualisation can lie, for one thing, in the qualitative and quantitative constancy of composition of the subject under study. In this sense we may speak of the identity of a portion of water and a piece of ice which is formed from it, or of the identity of ethyl alcohol and ether. It is obvious that science, treated as a set of mutually interconnected statements, does not preserve such an identity in the course of its evolution. On the contrary, it is this aspect that changes most.

Secondly, it may be possible to speak of a structural identity. In this case we note that although the individual components of the object under study may change, the system of relationships between its components remains the same. It is in this respect that an organism maintains its self-identity in spite of the metabolic process, or for that matter a whole biological species, whose individual members die and are born and even when their numbers may change, still the inter-individual relationships remain unchanged. In this respect also it would be difficult to speak of the self-identity of science in the process of its evolution for in this process not only the content of statements but also their mutual relationships undergo change.

In his work *The Direction of Time*, Hans Reichenbach has used the concept of *genidentity*. According to Reichenbach, if we are saying that an object at T_1 and T_2 is genetically the same, regardless of any differences between the two states, it means that there is a historical continuity between these states, that they are genetically connected with each other. It seems that behind the common intuition which compels us to see as *science* all that in different epochs passed under this name, we may find just this genetic continuity which is determined by the social inheritance of knowledge from generation to generation, by the stability of certain social institutions – schools, academies, universities, as well as of social groups which are involved in the acquisition and transmission of knowledge.

One can doubt, however, whether genetic continuity constitutes a sufficient condition in order to ascertain that we are dealing with the same object. If self-identity were to be determined solely on the basis of genetic continuity, then we would be forced to conclude that every object exists eternally, even though all of its features and functions might have changed in the process of evolution. In any case, this type of self-identity does not permit us in any way to identify any specific features of the object, which is precisely the difficulty that the historian of science must deal with. There is little help in remarking that not everything changes at the same time, that in every change from one state to another something is preserved, if after a specific period of time nothing remains that would be common to the starting point of the evolution. This remark is nothing more than a differently formulated belief in the occurrence of *genidentity*.

If, therefore, it turned out that in studying the history of science we should not be able to ascertain any other continuity apart from the genetic, we would be forced to conclude that there is little sense in using the term *science in general*, as it denotes simply an unordered series of different although genetically related phenomena. This, of course, would be equivalent to saying that there is no way to answer the question 'what is science?'. In consequence, we could only speak then of the science of particular periods, with the stipulation that for every period this term has different meanings.

VII

There is, however, another type of identity which we have not taken into consideration up to this point, namely *functional identity*. This type of identity can be said to occur when, regardless of any other changes the object may undergo, it performs some constant function within a whole to which it belongs. It might appear that functional identity implies in reality structural identity, assuming that if the structure of an object changes, it will cease to perform the same function. This objection might indeed carry weight if the larger whole to which the object belongs did not itself undergo any change. However, when the opposite is true, the changeability of structure not only does not come in conflict with the stability of function, but can be its necessary condition. If the component has to fulfill a permanent function in the frames of an evolving whole, its structure must change.

The question arises: is it possibly true that what is constant in the process of evolution of science is some of its functions in human culture? Should this hypothesis be true, it would not be hard to understand why the traditional philosophy of science, which treats its subject as an autonomous and isolated entity, cannot cope with the problem at hand. The more detailed the analysis it performs, the more it is unable to treat its subject as a whole. It sees science as a series of genetically connected phenomena but cannot specify their specific *differentia*. It seems that this difficulty is peculiar not only to the philosophy of science.

Knowledge fulfills two major roles in human society: a practical one and a philosophical one. Men have always had to build shelters, seek out food, cure illness, measure time and space, produce tools and utilize them. At the same time they have always wanted to know why they should proceed in that, rather than in another manner.

In consequence, we find in all cultures at least some elementary knowledge of geometry, arithmetics, the ability to construe calendars, some knowledge of 'chemistry', 'physics', 'biology', 'medicine', etc. Nobody will doubt that many particular disciplines of science were born under the impact of everyday problems, as those concerning measuring a field, architecture, the martial arts, the domestication of plants and animals, the practice of agriculture, metallurgy, dyeing of wool, medicine, or the regulation of social life, etc. However, even the most impressive and un-suspected practical skills and the knowledge which made them possible in primitive civilizations are not treated as science, for we believe that science is a system of 'organized', 'ordered', 'systematized' knowledge. What is the trouble is the fact that it is so hard to state, what this system-atization delimiting science consists in.[30] The difficulty springs, at least partly, from the changeability of this very systematization.

At the same time, in all cultures, we find certain systems of beliefs – myths, religions, philosophies – which are to provide answers to 'cosmo-logical' questions concerning the structure of the universe and man's place in the order of things, as well as to regulate his behavior. And not always were these beliefs expected to be coherent with the knowledge on which everyday practice was based. Thus, according to Egyptian cosmol-ogy, "the sun god, Ra, travels in his boat across the heavens each day, but there is nothing in Egyptian cosmology to explain either the regular recurrence of his journey or the seasonal variation of his boat's route."[31] It is obvious, however, that the ancient Egyptians knew that during certain periods the days are longer and nights shorter, while in others the opposite is the case and, moreover, that this cycle is regular. It appears that this cosmology was not expected to explain such 'details'. The cosmological system which sought to explain the world as a whole, and the knowledge applied in everyday activity, did not have to constitute a coherent system of beliefs, even when both of them were applied simultaneously. This is obvious in such cases as that of a priest utilizing salves and herbs to heal a sick man, while at the same time pronouncing a magic spell.

Thus, it seems that science came into being with the requirement of such coherence and that one of the functions it performs permanently in human culture consists in unifying into a coherent system practical skills and cosmological beliefs, the *episteme* and the *techne*. It is, of course, hard to pinpoint the place and the exact moment when this requirement was accepted for the first time, but it seems certain that the Greek science tried to fulfill it. At least from that time the history of science consisted, at least in part, in unifying these two spheres of knowledge with which science, with its feet on the ground and head in the skies, was permanently connected. Changing constantly under their impact, it changed them in turn. The changeability of the systematization and organization of scientific knowledge is a consequence of the function it performs, is a means of realizing this function in the face of changing human experience. With its feet solidly on the ground, science has always been 'empirical', but the sphere of human experience, which it was to incorporate into a cosmological order, has not always been identical. With its head in the skies, science, however, has never been purely empirical; not only because it always has to supplement empirical data by 'speculations', but also because it has to accomplish this function by coordinating two spheres of knowledge, the knowledge of 'how?' and 'why?', of what is practically possible, and what is theoretically impossible, of the universe of human fortuitousness and of cosmic necessity. It cannot be purely empirical because it has to coordinate the discovery of 'facts' with formulating general principles which do not follow logically from these facts. From this point of view, the efforts of Aristotle, Copernicus, Descartes, Newton and Einstein were directed towards the same goal and fulfilled the same function in culture at different stages of its evolution.

By stating that during its history science has fulfilled this function, I am not suggesting that this was its sole function nor that its relationships with these two spheres of knowledge, which it has to unify, have been constant. There have been periods in the development of science when the main stress was laid on *techne* and, also, periods when it was laid on *episteme*. What I do suggest is that, despite all changes that science might have undergone, this is its permanent and specific function which differentiates it from other products of human intellectual activity. Any cognitive effort which does not attempt to fulfill this function can only be called pre-scientific. On the other hand, any cognitive effort, no matter how it is systematized and organized, which – whether under the pretext of specialization, of practical needs, or of avoiding metaphysical speculation – programatically renounces fulfilling this function, ceases to be science; it limits itself to producing technological prescriptions for the manipulation of objects which happened to enter into the sphere of human experience.

The claim that science has always fulfilled this function is not at odds with the obvious fact that during different periods it was performing different social roles and that its place and status in human culture were not permanent. The forementioned function is not the unique one it performs, while its place and status in human culture depend not only on these functions, but also on how they are evaluated,[32] and, thereby, on how scientific knowledge is applied.

Although it is undoubtedly true that since the times of Bacon and the Royal Society science has been valued and practiced more and more as a consequence of the power it bestows and can bestow, and that its social function has been profoundly affected by these circumstances, nevertheless this does not mean that it has ceased to fulfill the function we have singled out above. This is attested by the scientific activity of such modern and contemporary scientists as Newton and Einstein, Laplace and Hoyle, Maxwell and Bohr, Darwin and Morgan, Sherrington, Pavlov or Wiener. If sometimes we tend to lose sight of these facts and see science merely as a system of rules of efficient action and, by overestimating this function, to project such a vision of science into the past, it is a result of a specific social situation in which science has been split[33] into a provider of true knowledge on the one hand, and a provider of consumer goods on the other; as well as of the attitude which regards "technological efficiency... as the highest value".[34] By the very same token we lose sight of the fact that the genetic continuity of science has been rooted in its constant attempt to introduce such an order into the realm of knowledge which, in any given stage of human experience, will render possible the unity of practical action with man's global view of himself and of the world around him. The limitation of the interests of science to one of these spheres would of necessity constitute a break in this genetic continuity.

Every systematization of knowledge which fulfills this function, we have come to term a 'rational' one. In this sense, we may assume that the history of science is a series of attempts at a rational organization of knowledge of 'why?' and 'how?', and that the methodological criteria on which the science of any given historical period is based are subordinated to the contemporary understanding of this rationality.[35]

Thus we have two visions of science: the first aims at constructing normative, supra-historical criteria of demarcation, treats them as criteria of rationality and *ex definitione* excludes from the sphere of science any and all statements which do not satisfy these criteria; the second singles out science as a particular fragment of human culture, which performs the function of unifying the changing universe of human experience with no less fluctuating ideas concerning the cosmological order and human cognitive capabilities, and treats the changing methodological rules as means for accomplishing this function. These two different visions of science must lead to different programs of philosophical reflection on science and on its evolution.

NOTES

[1] K. R. Popper, *The Logic of Scientific Discovery*, London 1956, p. 34.
[2] Cf. R. Carnap: 'Ueberwindung der Metaphysik durch Logische Analyse der Sprache', *Erkenntnis* 2 (1932). [English translation in *Logical Positivism*, ed. Ayer, Glencoe, Illinois, 1959, pp. 60–81. – Ed.]
[3] R. Carnap: 'Testability and Meaning', *Philosophy of Science* 3 (1936) and 4 (1937).
[4] H. Reichenbach, *Experience and Prediction*, Chicago 1938; *The Rise of Scientific Philosophy*, Berkeley and Los Angeles, 1951.
[5] C. G. Hempel, 'The Theoretician's Dilemma', *Minnesota Studies in the Philosophy of Science*, Vol. II, 1958; 'Problems and Changes in the Empirical Criterion of Meaning' *Revue Internationale de Philosophie* 4 (1950), 41–63.
[6] R. Carnap: 'The Methodological Character of Theoretical Concepts', *Minnesota Studies in the Philosophy of Science*, Vol. I, 1956.
[7] K. R. Popper, *op. cit*; *Conjectures and Refutations*, London 1969.
[8] Cf. below, Chapter 5.
[9] W. V. O. Quine, 'Two Dogmas of Empiricism', *From a Logical Point of View*, New York 1961.
[10] H. Mehlberg, *The Reach of Science*; Toronto 1959.
[11] L. Kołakowski, *The Alienation of Reason*, New York 1968, p. 206.

[12] Popper clearly stresses the normative character of his criterion of demarcation, "... what I call 'methodology' should not be taken for an empirical science ... And my doubts increase when I remember that what is to be called a 'science' and who is to be called a 'scientist' must always remain a matter of convention or decision." *L.S.D.*, p. 52; cf. *ibid.* pp. 9, 10, 11.

[13] R. Ingarden wrote: "In the lecture given to the VIII International Congress of Philosophy in Prague, I tried to demonstrate that the neopositivist point of view cannot withstand criticism because the statements which are to belong to philosophy, i.e., metalogical statements should have been meaningless according to the basic assumptions of neopositivism", *Z badań nad filozofią współczesną*. (*Studies on Contemporary Philosophy*) Warsaw 1963, p. 651. (Cf. also pp. 655–662.)

[14] K. R. Popper, *Conjectures* ... p. 259.

[15] K. R. Popper, *Conjectures*, p. 259.

[16] K. R. Popper, *L.S.D.* pp. 52–53.

[17] K. Ajdukiewicz, *Logika Pragmatyczna* (*Pragmatic Logic*), Warsaw 1965, pp. 173–177.

[18] *Ibid.* p. 175.

[19] K. R. Popper, *L.S.D.* p. 55.

[20] I. Lakatos: 'History and its Rational Reconstructions', in *Boston Studies in the Philosophy of Science*, Vol. VIII (ed. by R. S. Cohen and R. C. Buck), Dordrecht 1972, p. 91.

[21] W. V. O. Quine, *op. cit.*

[22] *Ibid.* pp. 42–43.

[23] Cf. E. Cassirer, *Substance and Function*, New York 1953, Part I.

[24] K. Mannheim, *Ideology and Utopia*, New York 1936, pp. 259–261.

[25] T. S. Kuhn, *The Structure of Scientific Revolutions*, Chicago 1970, p. 2.

[26] *Ibid.*, p. 3.

[27] L. Geymonat, *Filosofia e filosofia della scienza*, Milan 1960.

[28] T. S. Kuhn, *ibid.* pp. 2–3.

[29] L. Geymonat, *ibid.*, pp. 26–27.

[30] Cf. E. Nagel, *The Structure of Science*, New York 1961, Chapter I.

[31] T. S. Kuhn, *The Copernican Revolution*, p. 7.

[32] Cf. my 'Nauka i wartości', '*Zagadnienia Naukoznawstwa* 1 (1971), and 'Scjentyzm i rewolucja naukowo-techniczna', *Zagadnienia Naukoznawstwa*, 3 (1970).

[33] Cf. K. Pomian: 'Działanie i sumienie', *Studia Filozoficzne* 3 (1967); S. Amsterdamski, 'Nauka współczesna i wartości, '*Zagadnienia Naukoznawstwa* 1 (1971).

[34] L. Kołakowski, *op. cit.*, p. 202.

[35] Cf. W. W. Bartley, 'Theories of Demarcation between Science and Metaphysics', in *Problems in the Philosophy of Science*, Vol. 3, Amsterdam 1968, pp. 40–64, and the discussion on this paper, pp. 64–119.

CHAPTER III

THE CONTEXT OF DISCOVERY AND
THE CONTEXT OF JUSTIFICATION

I

In a stimulating paper of S. Rainko, we read:

We can analyze knowledge either as a static system and concentrate on studies of its language, its structure and the procedures which it assumes (these are the problems mainly investigated by the contemporary methodology of science) or as a dynamic, continually evolving system whose development is regulated by certain rules. The first point of view could be called synchronic, the second, diachronic. Accordingly, one may speak of a synchronic and of a diachronic epistemology or methodology. The traditional history of science was not able to provide a theory of evolution of knowledge and remained an ideographic discipline with its own specific field of interests. It seems that even in view of the changes that history of science underwent in recent times, it cannot replace the methodological analysis of problems posed by evolution of knowledge. For this we need a separate discipline, but one that would remain in close contact with history of science.[1]

This statement is in full agreement with what we have said above. The "separate discipline that would remain in close contact with history of science" (and – we might add – with the sociology of knowledge) is just what we have earlier called the *philosophy of science*.

One qualification is needed, however. It is not quite true that the contemporary methodology provides only synchronic analyses and that it is only this attitude which is responsible for the incompatibility of visions of science provided by philosophy on the one hand, and history on the other. It is not enough to undertake diachronic studies on science in order to overcome these incompatibilities. The problem is how these diachronic studies should be conducted. And it is just this problem we will deal with in this chapter.

II

The logical empiricism of the thirties of this century is, of course, the best known example of a synchronic methodology. It was interested exclusively in the logical structure of acquired knowledge, not with the problems of its development. (It is just this epistemology Rainko had in mind when he stated that contemporary methodology limits itself to synchronic analyses.) Within such a framework, historical questions could not even be asked and the scientific achievements of the past could only be a subject of an a-historical criticism which saw them as not conforming with the criteria of demarcation, meaningfulness or rationality assumed by this philosophy. These achievements could be only a subject of criticism because it appeared (and it could not be otherwise) that scientists of previous epochs, biased by myths, superstitions or metaphysical beliefs, did not proceed as they ought to, and that the theories they constructed did not fulfill the rules of scientific method postulated by this philosophy. What is more, since the proposed rules of scientific method were treated as supra-historical standards of rationality, the history of 'genuine' science could only begin with the very latest revolution in the development of knowledge. Drawing the extreme consequence from this opinion, we would be obliged to state that science has no history at all. What is considered as its history is simply a series of mistakes, misunderstandings and opinions which, from the point of view of this philosophy, cannot be treated as scientific. The problem of reconstruction of the process of growth of knowledge was beyond the interests of this philosophy.[2]

It is well known that this attitude was much criticized, especially by those historians of science who were convinced that such a science as presented by logical empiricism never actually existed, who aspired to overcome the purely ideographic character of history of science and were, therefore, sensitive to the opinions of philosophers. Within the 'professional' philosophy of science, it was Popper who revindicated the problems of growth of scientific knowledge during the thirties of this century.

In the *Introduction* to the first English edition of his *Logik der Forschung* (originally published in 1934; the main ideas having been already formulated in the never published essay: *Die beide Grundproblemen der Erkenntnistheorie*), Popper wrote as follows:

The central problem of epistemology has always been and still is the problem of the growth of knowledge. *And the growth of knowledge can be studied best by studying the growth of scientific knowledge.* I do not think that the study of the growth of knowledge can be replaced by the study of linguistic usages, or of language systems.[3]

The whole preface, as well as the book itself, is full of polemics with that conception of philosophy of science which limits itself to the analysis of

CONTEXTS OF DISCOVERY AND JUSTIFICATION   49

the logical structure of acquired knowledge, i.e., with synchronic epistemology. What really interests Popper is not the logical structure of science, but rather the logical and methodological rules due to which science grows, and the old theories are replaced by new ones. He wants to reconstruct the logical pattern of the process of evolution of knowledge and specify the methodological rules whose application would be the most favorable for this process.

There is, however, one point which is common to both Popper's program and that of logical empiricism. Both of them claim that the philosophy of science is to limit itself to the study of logical problems, i.e., to the *context of justification*. In the case of logical empiricism this is to be the study of the logical structure of acquired knowledge, while in the Popperian program, it is the study of the logic of its development, the logic of scientific discoveries, which is to serve as a basis for the reconstruction of the process of growth of knowledge. (As we will see later, this point of view is perhaps the most permanent component of Popper's reflections on science from the time he published his *Logik der Forschung* until his last book *Objective Knowledge*, and it has had tremendous impact upon his philosophy. It is closely connected with his anti-psychologistic and anti-historicist attitude, with his opinion about the priority of logical over genetical – i.e., historical, sociological or psychological – questions.)

Thus, taking into account the difference between synchronic and diachronic epistemology as pointed out by Rainko, we must state that Popperian falsificationism must be classified with the latter. This qualification is not simply an academic question. What we want to stress here is not only the important role of Popper's thought in the development of that trend in the philosophy of science which accepts the view that the theory of knowledge should be in agreement with facts provided by the history of science, but also that it is much more important to realize that although within the framework of logical empiricism, the limitation of interests of philosophy of science to the context of justification is not in contradiction with the goals which this methodology aims at, yet in the framework of a program which seeks to reconstruct the process of growth of knowledge, its mechanism, this delimitation, leads to essential difficulties.[4] This brings us back to what we stated above, that it is not enough to undertake diachronic studies; the question is how they should be conducted.

This question is obviously connected with the problem of demarcation, as we discussed it earlier. What is common to all the attempts to solve the problem of demarcation is an understanding of science as a system of those statements which are susceptible to the verdict of experiment. And when the growth of knowledge is seen as the evolution of such a system, there is nothing left to do but to study the logical relationships between the components of the system. The development of knowledge is seen as a process which is determined solely by relationships between different claims of science, that is, among those statements which satisfy the criterion of demarcation. Science is treated as an autonomous element of human culture and one disregards, or at least one does not take into account, other factors of its development, considering them as irrelevant.

It is important, therefore, to reflect upon the question of how justified it is to restrict the philosophical reflection on science to the context of justification and what are the difficulties this restriction leads to.

III

In describing a widely accepted view on the subject of philosophy of science, Herbert Feigl wrote:

There is a fair measure of agreement today on how to conceive of *philosophy* of science as contrasted with the history, the psychology, or the sociology of science. All these disciplines are *about* science, but they are 'about' it in different ways.... In the widely accepted terminology of Hans Reichenbach, studies of this sort pertain to the *context of discovery*, whereas the analyses pursued by philosophers of science pertain to the *context of justification*. It is one thing to ask how we arrive at our scientific knowledge claims and what socio-cultural factors contribute to their acceptance or rejection; and it is another thing to ask what sort of evidence and what general, objective rules and standards govern the testing, the confirmation or disconfirmation and the acceptance or rejection of knowledge claims of science.[5]

A similar position is held by other authors; thus Popper, for example, writes:

The question how it happens that a new idea occurs to a man – whether it is a musical theme, a dramatic conflict, or a scientific theory – may be of great interest to empirical psychology; but it is irrelevant to the logical analysis of scientific knowledge. This latter is concerned not with *questions of fact* (Kant's *quid facti?*) but only with questions of *justification or validity* (Kant's *quid juris?*). Its questions are of the following kind. Can a statement be justified? And if so, how? Is it testable? Is it logically dependent on certain other statements? Or does it, perhaps, contradict them? In order that a statement may be logically examined in this way, it must already have been presented to us. Someone must have formulated it, and submitted it to logical examination.[6]

Although Popper does not use the term 'philosophy of science' but 'logical analysis of knowledge', there can be no doubt that he considers them as equivalent. In both cases, that is of the logical empiricists and Popper, the strict differentiation between the context of justification and the context of discovery is a result of the acceptance of Husserl's criticism of psychologism[7] in logic and, of course, of the restriction of philosophy to logical questions. Just as, according to Husserl, the laws of logic do not pertain to how one thinks, but how one should think in order to think correctly, so philosophy of science, reduced to logic and methodology, does not pertain to how science is done but how it should be done in order to be done rationally. It is not concerned with psychological motives or historical circumstances which drive us to accept a statement, but with logical rules which justify it.

It must be noted that the limitation of the philosophy of science to the context of justification and the exclusion from it of any genetic (historical or psychological) questions means, for example, that inductivism is limited to the thesis that inductive confirmation is the method of accepting scientific claims, whereas the problem of the method of arriving at these claims is not a philosophical (epistemological) problem, and philosophy of science is not concerned with it. So we might say that from this point of view, epistemology is limited to problems concerning the assessment and control of knowledge, while the problem of how knowledge is acquired is excluded from it. In his recent *Objective Knowledge* Popper pushes this conception to its extreme consequences, and defends 'an epistemology without a knowing subject'. He states that what is the object of philosophy of science is a 'third world' of thought contents and of critical arguments, something like Plato's world of ideas or Hegel's objective spirit, while subjective knowledge belongs to the 'second world' which philosophy of science is not interested in.[8]

There is, of course, no need to stress the fact that the question of the way of arriving at a statement, of its psychological or historical origins, is not equivalent with the question of its logical value or the degree of its confirmation. Answers to the questions 'how did you arrive at this belief?' and 'how can you justify this belief?' are not usually the same (except the case when the statement in question has been deductively derived from premises taken as granted). The neglect of this obvious difference may lead, and has often in the past led, to serious misunderstandings.

It is one thing, however, to distinguish between these two kinds of questions (the methodological and genetic) and quite another thing to assume that this difference should serve as a basis for the demarcation between different disciplines studying science, i.e., that the philosophy of science should be concerned only with methodological problems, while sociology, history or psychology, with the genetic ones. The assertion of this difference is not a sufficient condition for specifying what kind of questions are the legitimate subject for philosophical investigations.

Why, in fact, should we at all respect this demarcation line? Only as a convention, or rather because of some theoretical reasons; and if so, then for which one? In any case, the right observation that the genetic and methodological questions are of a different nature does not imply that philosophy of science is to deal only with the latter. For a theoretical justification of this opinion some additional premises must be assumed. These premises would have to state that in the process of evolution of knowledge there are no connections between the accepted method of justification of statements and the manner of arriving at them or, at least, that these connections are irrelevant for the understanding of what science is and how it develops. It is obvious that these connections are non-existent if, and only if, the methodological rules of accepting, refuting and testing scientific claims would be historically unchangeable, permanent.

What is more, these premises would have to state that the understanding of science (and of its growth) is nothing more than the understanding of its logical structure (or of the logic of its development, that is of methodological rules according to which old theories are replaced by the new ones), and that the history of science and the sociology of knowledge have nothing to contribute to this understanding. That indeed would be the case only if the growth of knowledge were determined exclusively by the methodological rules for accepting or refuting scientific theories.

IV

It does not make sense to deny anyone the right to limit his philosophical interests to the problems concerning the context of justification or even to reserve only for them the label of *philosophy of science*. If the well founded distinction between the context of justification and the context of discovery were to serve as a principle for establishing a purely conventional border-line between the domains of philosophy of science and other disciplines about science then, from the formal point of view, this distinction should not have any impact upon the solution of problems studied accordingly by these disciplines. Problems, which in conformity with this borderline would be classified as 'not philosophical', would be investigated by means of appropriate methods by other disciplines. And the mere fact of which one of these disciplines will provide the solution is of little importance in so far as the very content of this solution is not affected by the conventional decision.

The matter is not so simple, however. The conception of limiting the interests of philosophy of science to problems pertaining to the context of justification is not a conventional proposal for delimiting the domains of particular disciplines which study science. It is rather a theoretical thesis which assumes that in order to reach an understanding of science and of its development, it is in principle necessary to study the context of justification which is then considered to be absolutely independent from the context of discovery. When the limitation of philosophy of science to studies of problems pertaining to the context of justification has been sanctioned in this way, it becomes the basis for a postulate indicating not which problems are to be classified as the subject of these disciplines, but rather pointing out how these questions which (on the grounds of some as-sumptions or of tradition) are treated as philosophical, should be studied. It is evident that in this case the discussed limitation determines the methods of study and must have impact upon the solutions which will be reached.

If, for example, one assumes that the philosophy of science is to deal only with problems pertaining to the context of justification, and at the same time one assumes, as Popper does, that the problem of the growth of knowledge is the central problem of epistemology (and, by the same token, of the philosophy of science), then the study of the growth of knowledge must be limited to logical questions, that is to the *logic of scientific discovery* which is expected to provide the appropriate solutions. The point, however, is that the lack of logical connections between ques-tions pertaining to the context of justification and others pertaining to the context of discovery does not mean that there are no historical connections between them and that the process of the growth of knowledge may be explained without studying these connections.

The discussions and polemics which followed the publication of Kuhn's *The Structure of Scientific Revolutions* may serve as a good example of the real function of the discussed differentiation. The discussion centered on the Kuhnian thesis of the "insufficiency of methodological directives, by themselves, to dictate a unique substantive conclusion to many sorts of scientific questions"[9] which confront the scientist, and about his conclusion that in order to understand the evolution of scientific knowledge it is necessary to go beyond logic and methodology, to take into account the sociology and psychology of scientific communities. In order to understand the development of science it is imperative, according to Kuhn, to study the values which have impact upon scientists in the process of scientific research, as well as the institutions which organize the research. "... the explanation of [the mechanism of the evolution of knowledge] must, in the final analysis, be psychological or sociological. It must, that is, be a description of a value system, an ideology, together with an analysis of the institution through which that system is transmitted and enforced."[10]

This very thesis cannot be accepted by those who maintain that there is no reason to go beyond the logic of scientific discovery,[11] and that the context of discovery is irrelevant for the understanding of the process of growth of knowledge.

None of Kuhn's critics, however, have maintained that the problem which he undertook does not belong to the field of philosophy of science because it ranges beyond the context of justification. It was argued rather that in the philosophy of science there is no reason to go beyond the boundaries of methodology, that the proposed attitude leads to irrationalism or, more precisely, to the opinion that the development of science is not a purely rational process.

Now, no matter who is right in this discussion (we will come back to this problem later), it shows that the limitation of the philosophy of science to questions pertaining to the context of justification is not a matter of the formal boundaries between particular disciplines studying science, but of methods of study and must, therefore, profoundly affect the solutions of problems.

At the same time this polemic shows that the Popperian program of the logic of scientific discovery was fraught *in nuce* with serious difficulties. On the one hand, he introduced into the philosophy of science, which limited itself to questions pertaining to the context of justification, the diachronic problems of the growth of knowledge. On the other hand, he accepted one of the basic assumptions of this trend in philosophy and reduced the theory of the growth of knowledge to the *logic* of scientific discovery. We believe, and we will try to demonstrate, that this resulted in some strains in his system. Nevertheless. Popper never gave up this program. What is more, he believes that the abandonment of this program must result in some kind of irrationalism. That is why he is so critical toward all those continuations of his program which, stressing the importance of diachronic problems for philosophical reflection about science, gave up his assumption of the priority of logical questions over historical ones, the assumption which he considers as the *conditio sine qua non* of rationality. Certainly, among those authors who 'betrayed' this program by developing some of its ideas, are Kuhn, Feyerabend,[12] and Polanyi.[13] In this sense, perhaps paradoxically, and certainly against his own intentions, Popper is 'responsible' for inspiring a trend in the philosophy of science which does not lie well with his own convictions. His logic of scientific discovery opened, in effect, the way to the posing of many questions, which could not have been asked within the framework of synchronic epistemology, but could not satisfactorily solve them. The answers to these questions, provided by the aforementioned authors, are incompatible with the Popperian program of philosophy of science as the logical reconstruction of the process of growth of knowledge.

The aforementioned modifications – the conception of the objective existence of the 'third world' and the 'epistemology without a knowing subject' – which Popper introduced into his philosophy of science were just meant, as I believe, to defend his program against this criticism, even at the price of some kind of Hegelianism or Platonism. His logic of scientific discovery becomes here something like a logic of evolution of the 'third world' of objective knowledge existing autonomously of any knowing subject.

In order to avoid any misunderstandings we must add that we are not concerned here with the irrelevant question of how the various fields of studies of science should be labeled, but how their problems are to be specified. This is what decides their solutions.

We believe that the differentiation between questions pertaining to the context of justification and those of the context of discovery could provide a basis for the demarcation between philosophy of science and other disciplines about science, if these questions had not only different meanings (as they have) and were logically independent (as they are) but also *if in studying the process of the growth of knowledge we could neglect their historical connections*. Only under this condition would the suprahistorical logic of scientific discovery have been able to provide a satisfactory account of the process of the growth of knowledge. We do not believe, however, that this condition is satisfied in the real process of evolution of science.

V

What, however, do we mean by an 'adequate solution'? The appraisal depends, of course, largely on the accepted research program, that is, on our opinion about its tasks and on our expectations concerning its possibilities. A solution which is adequate from the point of view of one program may be inadequate in the frame of another program which aims at different goals. Thus, when we state that some solution is not satisfactory, we are obliged to specify whether it does not meet the requirements set by the program, or whether we do not accept the program as such. We may reject it, not because it cannot cope with problems it raises, but because we judge it to be too limited and neglectful of a number of problems which are important. This is the well-known distinction between immanent and external criticism.

In order to appraise the postulate restricting philosophical reflection on the development of science to the study of the context of justification (to the logic of scientific discovery) we have, first of all, to realize what the goals are which philosophy of science has to achieve according to the adherents of this program. Second, we have to answer the question whether the accepted postulate is able to fulfill its promises, and if not, why not. It is these two questions that we will discuss now.

It must be noted, at the outset, that the programmed limitation of philosophy of science to questions pertaining to the context of justification is intimately linked with a specific manner of articulating the subject under study. What constitutes this subject is science understood as a system of statements, accepted (either diachronically or synchronically) as true and fulfilling certain methodological requirements. In other words, it is science understood as a product of human cognition, objectivised in intersubjectively communicable and intersubjectively controlable sentences. The potential set of these statements is delimited by the accepted criterion of demarcation.

This view was perhaps most clearly formulated recently by Popper when he distinguished the third world, that of objective knowledge stated in sentences.[14] This attitude, however, is common to many contemporary works on methodology, although it is not always formulated as radically as by Popper. In the third of these worlds there does not, and cannot exist any relationship between the manner of arriving at any specific statement and the manner of justifying it. When one approaches science as a ready product of cognition, as an autonomous product of human intellectual activity, and one studies it disregarding the historically conditioned human subjects (since they belong to the first, and the second world, but not to the third), then it is obvious that the problem of the development of science is reduced to the question concerning the rules, according to which this 'third world' evolves. The history of the third world, although it is 'man-made', must be equivalent with the logic of its development. In such a context any program of the philosophy of science which would go beyond the study of the context of justification must *a priori* be considered as mistaken. For psychology, history, or sociology of science have nothing to say about this 'third world'. So, we arrive at a conception of philosophy without a knowing subject, and in practice, without epistemology.

Popper believes that it is the only way to avoid subjectivity and irrationalism in the theory of knowledge, which cannot be objective and rational as long as it is interested in the 'second world'. "Although personal subjective beliefs can always be described as irrational in some sense, the use of the objective knowledge shows that there need not be any Humean conflict here with rationality."[15] It appears that his conception of supra-historical criteria of rationality may be defended only by eliminating the knowing subject whose beliefs are always 'in some sense irrational' as conditioned by 'non-rational', historical or sociological factors.

VI

Taking into account the above-mentioned basic assumptions, the philosophy of science cannot extend beyond the study of either the logical

structure of knowledge or, if it is concerned with the problem of development, beyond the logic of scientific discovery. Obviously, no one really
believes that the actual course of the history of science proceeds strictly
according to a preset logical mold, even if he assumes that in the whole
history of science the same methodological rules apply. It becomes clear
that, by necessity, scientists in the past, as well at present, do not proceed
precisely in the manner that would be prescribed by a rational model
constructed by methodology. But a philosophy of science which undertakes the task of the logical reconstruction of knowledge does not aim at
the precise description of how science is created and how it develops. It
postulates how science should develop in order to be rational, that is in
order that the accumulation of true knowledge and the elimination of
mistakes were as effective as possible.[16]

So, if somebody believes that philosophy of science has to reconstruct
in the 'third world' the process of the development of scientific knowledge,
and that this reconstruction is to serve as a model of rational inquiry and
as a frame of reference for the appraisal of the history of science, then any
argument which would indicate that this reconstruction does not give an
adequate account of the real process of the development of knowledge,
would be totally *irrelevant*. Such an adequate description is not the aim of
the logical reconstruction, and, what is more, is by definition impossible.

One question, however, comes instantly to mind: what are the causes of
this incompatibility between the rational model, which methodology so
painfully tries to construct, and the actual development of history. The
answer to this question will determine whether the proposed model of
rational research procedures is realistic and not Utopian or, in other
words, whether it is possible to proceed according to it in the practice of
research or in the evaluation of already articulated theories.[17]

The simplest solution would be to note that the proposed model of the
development of knowledge is meant to be normative and, therefore, it
makes little sense to inquire about its relationship to reality. The fact that
scholars are directed in their research by certain norms of conduct, that is
by methodology, can lead to two different questions: the first, what *are*
the norms of their conduct and, the second, what *should* they be. The
answer to the first question would then be descriptive, to the second,
normative. In the first instance, we are dealing with a descriptive methodology which concerns the description of methods used in science, in the

second, with a normative one, postulating norms, which should be applied. The difference is analogous to the distinction between the description of moral behavior, and the proposal of an ethical code of behavior. If this is so, then the disparities between the normative model and reality have no bearing on the evaluation of the normative model.

In practice, however, the matter is more complicated and the above answer is not adequate. For, in fact, we rarely have to deal with a purely normative methodology. Scholars who construct normative methodology, and here Popper may serve as a good example, are usually careful to ensure not only that their methodological constructs lead to the realization of certain values, which are hidden behind the choice of a given method, but also that they have some relation to reality and do not constitute a completely arbitrary set of decrees. The best evidence is that they take into account the history of the development of knowledge, and do not reconstruct it in isolation from the actual history of science. If their model were purely normative, they would be fully justified in ignoring all the historical data. It might be stated that their research oscillates between the study of reality (that is the evolution of science) as it is and the formulation on this basis of normative methodological rules. As a result, the character of their propositions is often unclear, not only because of the language in which it is couched, but also because many features of the proposed model correspond to the real process of evolution of science. In this sense, it may be said that normative models contain certain descriptive elements, that some of them are 'more', while others 'less' descriptive. This provides a new dimension for the evaluation of these models.

Therefore, although between that which is postulated by the normative model and reality, between what 'should be' and what 'is', there exists a difference which is assumed in advance, still one has the right to ask of such a reconstruction of the development of knowledge where these differences come from and of what nature they are. Although all such reconstructions assume certain discrepancies between the models and reality, nevertheless it seems that in the case of some kinds of discrepancies it should be stated that the reconstruction has so little to do with reality that one may very well question its usefulness either as a model of procedure, or as a standard for evaluation. A norm which prescribes a physically impossible action is not logically false since norms are not sentences which can be judged as either true or false. It is, however, Utopian and, as such, may be rejected. That such an appraisal of the reconstruction of the development of science on the basis of the logic of scientific discovery is justified may be shown by quoting one of the partisans of such an approach:

In writing a historical case study one should, I think, adopt the following procedure: (1) one gives a rational reconstruction; (2) one tries to compare this rational reconstruction with actual history and to criticize both one's rational reconstruction for lack of historicity and the actual history for lack of 'rationality'.[18]

We, therefore, may claim that our question concerning the origins and the nature of discrepancies between a rational reconstruction and reality are fully justified.

To repeat again, what we are asking here is whether the discrepancies between the reconstruction of the development of knowledge and reality are caused by some factors that are external to the development of knowledge, irrelevant for our understanding of the process, or if they are caused by factors that are relevant for this understanding. On the solution to this problem depends the answer to the following question: can a model, construed in normative terms, fulfill its function; that is, can one use it as a guide in research and can fully articulated theories be judged on its basis? We may, in fact, formulate this question in a different manner: can we visualise a situation in which, at least in approximation, the process would proceed in the manner postulated by a normative model, that is a situation in which we could consider the model as an acceptable idealisation?

If the discrepancies between the normative model and reality are caused by outside factors which have nothing to do with the very process of the evolution of knowledge, just as, for example, the variations in the velocity of free falling bodies are dependent on factors other than those specified by the law of free fall, such as the resistance of the medium in which the fall occurs (these factors are, of course, consciously disregarded by the law), then there at least exists a possibility that the proposed methodology could serve as a model for the rules of procedure of scientists as well as a model for the evaluation of theories. That is, it could fulfill its normative function. The model would not then be theoretically Utopian, and at the same time would mirror, in spite of its normative character, certain essential features of the development of knowledge. In the situation we could say that the logic of the development of the 'third world' of objective knowledge, gives an account of the essential features of the development of knowledge; or, that the 'third world' is a correctly stated object of study for the philosophy of science.

If, however, it should appear that the discrepancies between the model and the actual process of the development of science result from internal, essential mechanisms of the process and not from outside perturbations, from its features which are always present and which the analysis of the 'third world' must by necessity omit, then the model must be viewed as Utopian. It could serve neither as a model of procedure nor give a proper account of the development of knowledge. We would then conclude that the 'third world' does not constitute an adequate model for the study of the development of scientific knowledge.

By accepting the point of view that the logic of scientific discovery can provide an adequate account of the development of knowledge, we take *eo ipso* the position that particular historical circumstances in which science is practiced such as contemporary philosophical trends, the social status of science, opinions as to the tasks of scientific research, etc., are actually outside factors which deform and interfere with the normal, rational development of knowledge which, were it not for these outside interferences, would proceed according to the path indicated by the logic of scientific discovery. In this case the history and sociology of science are *irrelevant* for the understanding of the development of knowledge. Irrelevant, at least in that they do not explain the main line of the development of science, which is explained by logic, and only confirmed by history, and in that they only explain accidental deviations from the main path. They are irrelevant in that they explain not what is normal and rational but, so to speak, what is pathological in its development. They account not for what conditions this development, but what decides about certain deformations, even if they are inevitable, as nobody considers science to be something developing in a vacuum, in a sterile environment. With such an understanding of the development of science, its history must of necessity be of an ideographic nature.

Accepting such a conception, one could genuinely hope that the rational model of the development of knowledge could effectively fulfill its normative functions. One could expect that it could indicate for scholars the path along which they should proceed, provide them with criteria of rationality, demarcation, definitions of science which could warn them against sidestepping from the main path of the development of knowledge. It could warn them against deviations caused by non-rational, external factors of the development of science. The limitation of the philosophy of science to the context of justification would only be justified if this vision of the development of science were to be adequate.

We have previously mentioned the polemics concerning Kuhn's book. We have mentioned that according to his critics, the logic of scientific discovery provides, or can provide, *an adequate solution* to the problem of the development of scientific knowledge, and that the introduction of any other factors for this purpose is unnecessary, and what is more leads to irrationalism. Apparently, according to their opinion, this is so because the logic of scientific discovery accounts for the essential, basic features of the process of evolution of knowledge, and the deviations of reality from this logical model are irrelevant to the basic character of the process under study. This must mean that the confrontation of this model with reality is not an unjustified procedure, since what, in any other case, would 'an adequate solution' signify? The entire argumentation of Kuhn's critics attests to the fact that this is how they understand this thesis. This is perfectly understandable, as in the opposite case there would be no substantial differences between their opinions. The thesis that the logic of scientific discovery does not adequately explain the development of knowledge would not be in conflict with the opinions of its defenders if the former thesis were of a purely descriptive character, describing *how science really develops* and the latter purely normative, prescribing *how science should develop*. Behind these two positions, however, hides a totally different vision of science and of the mechanisms of its development.

VII

The rules for the reconstruction of the history of science on the basis of the logic of scientific discovery[19] had been presented by Lakatos in his work 'History and Its Rational Reconstructions.' As we read there, "... philosophy of science provides normative methodologies in terms of which the historian reconstructs 'internal history' and thereby provides a rational explanation of the growth of objective knowledge."[20] The internal history of science is "... a history of events which are selected and inter-preted in a normative way."[21] It explains that which is rational in the evolution of knowledge. The rational reconstruction never exhausts the history as: "The history of science is always richer than its rational reconstruction,"[22] and, therefore, "any rational reconstruction of history needs to be supplemented by an empirical (socio-psychological) 'external history'."[23]

'External' history explains facts of history which cannot be included in the rational reconstruction based on the logic of scientific discovery, that which on the basis of that logic appears as 'irrational' or 'non-rational'. When the real history of events does not coincide with the rational reconstruction, external history explains the origins of these inconsistencies. However, and this is of paramount importance, *"the rational aspect of scientific growth is fully accounted for by one's logic of scientific discovery,"*[24] and *"external history is irrelevant for the understanding of science."*[25]

This conclusion is not altogether surprising. First, one states that the logic of scientific discovery delimits the criteria of rationality; second, that it constitutes the basis for the reconstruction of 'internal' history; and finally, that what cannot be subsumed under this reconstruction is irrelevant for the understanding of science because 'internal history' fully exhausts the rational aspects of the development of knowledge.

But what, in this context, is meant by the terms 'rational' and 'non-rational'? Nothing else, it appears but "fitting, or not fitting, as the case may be, into the accepted supra-historical model of the development of knowledge". Or, what comes out to be the same, that the development of knowledge is a process which is *sui generis* logical, and the task of the philosophy of science is to reconstruct this logic. In any case, rationality is understood here in an a-historical manner since only in this case can supra-historical methodological rules of the logic of scientific discovery provide a basis for the evaluation of what in the history of science was rational and what was non-rational.

Lakatos is undoubtedly correct when he states that any history of science is always written from the point of view of some philosophy. He is correct as it is precisely the philosophy of science which provides the definition of what is science, what can be considered within its domain and what cannot, and how its development proceeds. His own methodology of research programs, as we shall demonstrate in Chapter VI, confirms this thesis. His analysis of the effects of inductivism, conventionalism and falsificationism upon the vision of the history of science, both its 'internal' and its 'external' history, are very instructive in this respect.[26] It is apparent, however, that the criteria of rationality, which he introduces, are determined by the criterion of demarcation which he accepted on the basis of a particular philosophy of science.

He is undoubtedly right, when he observed that on the basis of different philosophies of science, the line between 'internal' and 'external' history of science will fall in different places. Those facts, which on the basis of one particular logic of scientific discovery may be accounted for by rational reconstruction, on the basis of another one will have to be explained by 'external history' as they are non-rational and are to be explained in its terms.

It is obvious that he would like to present such methodological rules, that the reconstruction of history based on them would encompass the largest possible amount of facts from the history of science. He formulates even some meta-methodological criteria for appraising different reconstructions in this respect.[27] At the same time, however, the real history of science cannot serve as a test of the proposed reconstruction since the acceptable boundary of its criticism for 'a-historicity' is constituted by the accepted concept of supra-historical criteria of rationality. Therefore, facts which do not fit into this reconstruction must *ex definitione* be assumed as 'non-rational'.

As a result, if we agree that the task of the philosophy of science is the logical reconstruction of knowledge, since logic provides an account of the rational aspect of its development, then any discrepancy between this reconstruction and reality does not disqualify the reconstruction. Historians may eventually tell us *ex post facto* why such discrepancies had occurred, although in view of the rational development of knowledge they should never have happened. Facts, which on the basis of a given logic of scientific discovery cannot be explained as a link in the logical process, become relegated to the garbage dump of 'external' history which is irrelevant for the understanding of the development of science. The delimitation of the realm of problems of philosophy of science to the investigation of the context of justification and the treatment of the process of scientific cognition as independent from historical conditions, must lead either to the complete disregard for the actual history of science, or to the division of the real history of science into two parts. The first part would consist then of rational history, which follows the model of the logical development of science, while the second part is 'external', irrational history. It is irrational precisely because it does not fit into the model of logical reconstruction and it is irrelevant for the understanding of science.

Moreover, the reconstruction of the 'internal', rational history of science, spawned by this philosophy, cannot be *ex definitione* anything else but the history of ideas, expounded *modo geometrico*. When Lakatos asserted that whoever does not recognize the possibility of the reconstruction of the development of knowledge on the basis of the logic of scientific discovery is forced into irrationalism, it might also be said that he, attempting to fit the history of science into the scheme of the logical development of the 'third world', oscillates towards Hegelianism. Facts which do not conform to the constructed logical scheme expressing the rational spirit of the development of science are taken to be irrational and irrelevant for the understanding of the real process of the development of knowledge, which is in reality a realization of some supra-historical logic of scientific discovery; the discovery of this logic is the aim of the philosopher.

The line that he has drawn between 'internal' and 'external' history means that rational history runs always according to the same line of development and it winds its way through the external obstacles of fate. If it were not for these external complications it would proceed precisely by the route which is set out by the logic of scientific discovery and the 'internal history' which it reconstructs. 'External history' in this fortunate example would not exist at all. From such a point of view, any other determinations except the logical ones are indeed irrelevant for the understanding of science.

If, however, we treat science as a part of the intellectual culture of a given historical period, then the division of history into 'internal' and 'external' appears totally artificial and arbitrary. The essence of the development of science is then determined not only by its logic, but also by elements and factors which lie outside the sphere of logic. What from Lakatos' perspective appears as pathological, unnatural and irrational, from such a perspective appears as natural. Metaphysics, myths or superstitions are in some manner as immanently a part of science as the facts which we attempt to include into the rational reconstruction. The neoplatonic metaphysics of Kepler or Copernicus were as much an element of the rational organization of the universe which they attempted to reconstruct as the strictly empirical statements of their astronomic systems. If we try to understand the mechanism of the development of knowledge, the elements mentioned above cannot be analysed separately, one from another. What is more, the interdependence of these elements is by no means a *differentia specificam* of some particular period in the history of science; it constitutes one of the inseparable characteristics of science. To phrase it in a somewhat provocative manner, the history of science encompasses also the 'natural history of nonsense', especially when that nonsense appeared in the guise of sense as an element of a rational order through which the universe of experience was then viewed.

If we understand science in this manner, then the logical model of the development of knowledge not only cannot *constitute an adequate* description of the real history, which as we noted it does not pretend to do, but also cannot be utilized as a justified reference for the critical appraisal of any research procedures and their results. The view that any logic of scientific discovery can constitute a supra-historical model of scientific procedure, or for that matter a model of its evaluation is simply Utopian. It is Utopian because these procedures are determined by the function which science fulfills within culture and not by the changing tools with which this function is realized. There exists an alternative: either one constructs supra-historical models of rationality and then, inescapably, one finds irrational elements in the history of science (and nothing is changed by labeling them 'external' history, for at the most this provides one with the feeling of one's own rationality); or one attempts to see these 'irrational' elements as factors of a different order with the help of which science then, as today, attempted to rationalize the universe of human experience. However, to rationalize is not simple to describe. To rationalize one must go outside the world of experience and take recourse to principles which are to explain that experience. Therefore, science consists not only of statements about the universe under study, but also of assumptions about the knowing subject. The logic of scientific discovery which attempts to be an epistemology without a knowing subject by definition disregards this circumstance.

NOTES

[1] S. Rainko, 'Epistemiologia diachroniczna, zarys problematyki' *Studia Filozoficzne*, No. 48, 1967, p. 3.

[2] As Lakatos rightly stated: "As an epistemological programme it has been degenerating for a long time; as a historiographical programme it never even started." 'History and its Rational Reconstructions', p. 128, n. 70.

[3] K. R. Popper, *L.S.D.*, pp. 15–16. See also Popper, *Conjectures and Refutations*, p. 28.

[4] "Why did the logical empiricists have no interest in the logic of discovery? The historian of thought may explain this in the following way. Neoclassical empiricism replaced the old idol of classical empiricism – *certainty* – by the new idol of exactness. But one cannot describe the growth of knowledge, the logic of discovery, in 'exact' terms, one cannot put it in formulae: it has, therefore, been branded a largely 'irrational' process; only its completed (and 'formalised') product that can be *judged* rationally. But these 'irrational' processes are a matter for history or psychology; *there is no such thing as a 'scientific' logic of discovery.*" I. Lakatos, 'Changes in the Problem of Inductive Logic' in Lakatos, (ed.) *The Problem of Inductive Logic*, Amsterdam 1968, p. 328. And further: "...the Carnapians concentrating mostly on a rational 'synchronic' reconstruction of science and the Popperians remaining mostly interested in the 'diachronic' growth of science... I think the lack of recognition of this interdependence is an important shortcoming of logical empiricism in general and Carnap's confirmation theory in particular." *Ibid.*, p. 329ff. It might be added here that Lakatos shares Popper's opinion of a logic of scientific discovery which could account for the evolution of knowledge and which would be limited to the context of discovery if possible. Although I accept the criticism of logical empiricism as limited to synchronic problems, I cannot, however, accept this opinion on the possibility of conducting a diachronic epistemiology limited to the context of discovery.

[5] H. Feigl, 'Philosophy of Science' in R. M. Chisholm *et al.* (eds.), *Philosophy*, Englewood Cliffs, N.J. 1965, p. 472.

[6] K. R. Popper, *L.S.D.*, p. 31.

[7] The basic difference between the Husserlian understanding of the laws of logic as belonging to the universe of ideal relations and the neopositivist position, which understands them as linguistic rules, is irrelevant at this point. The antipsychologism of both positions is self-evident.

[8] We will have the occasion later on in this study to note that Popper's limitation of the philosophy of science to logical problems forces him to introduce into his methodology a number of conventionalist theses.

[9] T. S. Kuhn, *The Structure of Scientific Revolutions*, p. 3.

[10] T. S. Kuhn, 'Logic of Discovery or Psychology of Research' in I. Lakatos (ed.), *Criticism and the Growth of Knowledge*, p. 21.

[11] See K. R. Popper, 'Normal Science and Its Dangers', in *Criticism and the Growth of Knowledge*, pp. 51–59 and I. Lakatos 'Falsification and the Methodology of Scientific Research Programmes', *ibid.*, pp. 91–197, and in particular pp. 92–93.

[12] P. K. Feyerabend, 'Problems of Empiricism' in Colodny (ed.), *Beyond the Edge of Certainty*, Englewood Cliffs 1965, pp. 145–260, 'Explanation, Reduction and Empiricism' in *Minnesota Studies in the Philosophy of Science*, Vol. III, 1962, pp. 28–29 and 'Consolation for a Specialist' in *Criticism and the Growth of Knowledge*, pp. 197–231.

[13] M. Polanyi, *Personal Knowledge*, London 1958; *Knowing and Being*, London 1969.

[14] K. R. Popper, 'Epistemology without a Knowing Subject' in *Methodology and*

Philosophy of Science, Amsterdam, 1968, pp. 333–373. It is worth noting that such an understanding of science, as a system of formulated statements, differs from the position which sees it as a set of all possible statements which may be formulated on the basis of a deductive system irrespective of whether they have ever been formulated. (See K. Ajdukiewicz, 'Metanauka i metodologia' [Metascience and Methodology] in *Język i poznanie* [[*Language and Cognition*]]), Vol. II, Warsaw 1965, pp. 117–126.

[15] *Ibid.*, p. 81.

[16] "But this reconstruction would not describe these processes as they actually happen: it can give only a logical skeleton of the procedure of testing. Still, this is perhaps all that is meant by those who speak of a 'rational reconstruction' of the way in which we gain knowledge." K. R. Popper, *L.S.D.*, pp. 31–32.

[17] In distinction to Popper, who clearly claimed that the aim of the philosophy of science is to provide researchers with a rational model of procedure, Lakatos states that the basic aim of methodology is to provide not models of procedure, but rules for the evaluation of already articulated theories ('History and its Rational Reconstructions', p. 92).

[18] I. Lakatos, 'Falsification and the Methodology of Scientific Programmes', in *Criticism and the Growth of Knowledge*, p. 138.

[19] The attempt at constructing a 'historiosophy' of science on the basis of Popper's logic of scientific discovery is the work of his students – Agassi and Lakatos. I am not aware of any text of Popper's where he takes any position, negative or positive, towards these attempts.

[20] I. Lakatos, 'History and its Rational Reconstructions', *Boston Studies in the Philosophy of Science*, Vol. VIII, Dordrecht 1972, p. 91.

[21] *Ibid*, p. 108.

[22] *Ibid.*, p. 105.

[23] *Ibid.*, p. 91.

[24] *Ibid.*, p. 106 (my italics – S. A.).

[25] *Ibid.* ibid. p. 92 (my italics – S.A.).

[26] *Ibid.*, Chapter 7: 'Rival Methodologies of Science'. see also J. Agassi, *Towards an Historiography of Science*, The Hague 1963, where an analysis of the influence of inductionism on the historiography of science may be found.

[27] I. Lakatos, *op. cit.*, pp. 108–122.

CHAPTER IV

FACTS AND THEORIES: RADICAL EMPIRICISM

I

At the root of the controversy, which has been discussed in the two last chapters, lies hidden the problem of changes in the content of knowledge; that is, the problem of how the transition from old to new ideas occurs, and what is the essence of this process. In the case of the development of scientific knowledge, this problem concerns the replacement of old theories by new ones. Attempts to solve this problem belong to the traditional tasks of epistemology, while the solutions depend upon opinions concerning the relationship between facts and theories.

Among the different philosophies which have attempted to solve this problem, it is empiricism which particularly deserves our attention at this point. However, it is not so easy, as it might seem to be, to explain what the empiricist position is. There are as many answers to this question as there are authors writing on the subject. Furthermore, historians of philosophy labeled as *empiricism* a wide range of philosophical doctrines; from that of Aristotle through Bacon, Berkeley, Hume and Mach, to logical empiricism.

First of all, let us notice that empiricism in its various formulations was never an experimental theory which accounted for the process of the development of knowledge. It was a philosophical doctrine, in the light of which such theories were formed more or less successfully within the domains of various particular disciplines such as psychology, neurophysiology or information theory. By stating that empiricism is a philosophical doctrine, I have in mind primarily that it expresses a certain conception of man as a conscious being, i.e., of a being who, by his very nature, is capable of feeling, experiencing and communicating his psychological states and convictions to other men. By such a definition, empiricism would then be a specific ontological hypothesis concerning the relationship between man, considered as a psychical subject, and the universe. This hypothesis assumes that the senses are the only channel of communication between the knowing subject and the world and, therefore, the sense-experience is the necessary factor in every act of acquisition and control of knowledge. One may say that the core of empiricist philosophy is the understanding of man as a sort of measuring instrument which reacts by its feelers (the senses) to the states of the outside world as well as to its own states, and the assumption that there exists some kind of coordination between these reactions and the states or events that the 'instrument' reacts to.

On the basis of such an ontological hypothesis it is possible to build various theories concerning the 'instrument', i.e., different epistemologies. These must provide answers to such questions as: to what signals is the instrument sensitive? What is its 'resolving power'? Does it transform in any way the signals it receives and, if so, in what manner? In what way do the internal states of the 'instrument', such as memory, as well as the interaction with other similar 'instruments' affect its functioning?

The history of empiricism is the history of successive answers to such, and similar, questions. These answers (and this point is especially important) were largely, if not exclusively, formulated on the grounds of the analysis of the products of human intellectual activity. Before the rise of empirical psychology and neurophysiology it was impossible to study directly human cognitive processes. Therefore, the actual state and the scope of human knowledge has had a direct effect on the successive epistemological articulations of this ontological hypothesis.

Thus, for example, the scientific revolution of the 17th century, by introducing into scientific practice the use of measuring instruments (here this term is used of course in the literal sense) which mediated the relation between the knowing subject and the object under study, changed not only the scope of natural phenomena which could be subjected to scientific studies, but also the very concept of experience. At least since that time no epistemology could dispense with the concept of indirect experience and pass over all the problems it has introduced. It seems that it is rather this change and its consequences and not the alleged transition from speculative to empirical thinking which, along with the introduction of mathematical analysis of experimental data, constituted the real intellectual revolution which took place in that century.

The changes of the very character of experiment as a link in the process of cognition could not but change also the sense of the empiricist point of view. One may detect these differences by comparing the empiricism of Aristotle with the modern empiricism of Bacon, Locke, Hume or Mach. It would be enough to remember that according to Aristotle's empiricism, which constituted the very basis of his physics, a physical object was to be characterized above all by those features due to which it could be an object of direct sense-perception, due to which it could be touched, seen or smelled, etc. And the physical laws, accordingly, were to connect phenomena which could be perceived in direct contact with reality.

It would be difficult to overestimate the role which this concept of direct experience had to play in the intellectual revolution of the 17th century and the difficulties that it then raised.

In the same way, the difficulties faced by the empiricist epistemology in our times spring from recent changes in the concept of experience resulting from the advancement of knowledge. This progress has discredited, for example, the concept of an 'ideal observer', capable of acquiring information about an object by means of a measuring instrument without changing the state of the object. The first assult on this concept, which was the ideal of modern science, came as a result of the proved impossibility of 'Maxwell's demon'; the final blow was provided by Heisenberg's principle of indeterminacy. What is crucial for the understanding of an experiment as we know it is the fact that the measuring instrument is not simply a passive extension of our senses, but that it mediates some energetic and informational relationships between the knowing subject and the object. Any act of experience, even that which previously was called a passive observation, disturbs to some degree the state of the object and the information we receive about this state is at the same time information about the disturbance which the act of observation has caused, regardless of whether it was a direct or indirect observation. It is obvious that this fact must radically affect the way in which we understand such a concept as 'experimental data'.

Therefore, even if the basic premise of empiricism is cogent, there still remains the problem of its epistemological articulation which must provide an answer to the question, what does it really mean that experience is a necessary mediator in any act of acquisition and of control of knowledge?

Within modern, as well as contemporary philosophy, an important role has been played by a particular variant of the empiricist position. According to it, the idea that senses are a necessary mediator between the mind and the external world has been linked with the belief that the mind is simply a passive witness and registrar of experience, that sense-perception is an autonomous factor of cognition.

According to certain authors, the denial of this thesis constitutes the *specificam differentiam* of the anti-empiricist point of view. In order to justify their opinion they can, and often do, refer to a long philosophical tradition which in modern times can be traced back at least to Descartes and Kant.

On the one hand, the idea that the claims of science are not achieved exclusively due to the coercing power of sense-data, but that scientific facts are also products of theoretical thinking, had its roots in the philosophies of apriorism and conventionalism. We find it not only in the philosophy of Kant and his successors but also, undoubtedly under their influence, in the writings of Poincaré, Duhem, Le Roy and Popper.

On the other hand, logical empiricism, in continuing the trend of radical empiricism of Bacon, Locke, Hume and Mach, has rejected this conception. It put forward a program of logical reconstruction of science which has to be accomplished on the grounds of a purely empirical basis. This basis, interpreted either according to physicalism or phenomenalism, was conceived as free of any theoretical components. It was to consist of such statements that could be proved true or false only by means of experimental data and of rules of logic which, due to their analytical character, do not contain any information about the empirical world.

In spite of these widespread opinions concerning the difference between the empiricist and the anti-empiricist philosophy, I neither believe (what seems obvious) that the thesis concerning the impact of theoretical thinking upon the way of seeing scientific facts should be rejected only because of its philosophical genealogy, nor do I believe (what is not obvious and must be argued) that the content of this very thesis must always be, in all its interpretations, incompatible with an empiricist epistemology.

As I will try to argue, the claim that scientific facts are determined by empirical data as well as by theoretical thinking does not have to be equivalent with either Kantian apriorism or conventionalism. However, before undertaking this argument, let us analyze in more detail the point of view of radical empiricism, which perceives only the impact of facts upon theories and neglects the opposite relation between them.

II

The first and the most important consequence of the radical empiricist position is the absolutisation of facts, which is no better than the absolutisation of theories always denounced by the radical empiricist philosophy. According to this point of view, a fact, when once established, must remain a fact, both in the sense that every succeeding theory must explain it and, in the sense that statements accounting for this fact do not change their meaning in the frameworks of successive theories. This seems quite obvious: provided that experience is not only the unique, but also an autonomous source of all knowledge and the sole method of its control, nothing in principle (beside the removal of individual mistakes and the perfection of scientific instruments) can change an established fact. The sense of a statement which accounts for it is determined exclusively by observation and experiments. It means that *the theoretical mind deals directly with coercing empirical facts, while a theory is a superstructure which the mind builds up on foundations which are totally independent of it.*

Therefore, the relationships between facts and theories are considered as one-sided and 'non-reflexive'. Facts are invariants of our knowledge and constitute its final, unquestionable basis. The statements accounting for them (we shall refer to them as to the *observational statements*) are, by the same token, considered as a singular part of our knowledge which needs no further justification; experience itself compels us to accept or to refute them. We shall refer to this point of view as *inductivism*, noting that in some cases the thesis, stating that all our knowledge is based on observational statements, may be understood either in a genetic or in a methodological sense. In the first instance it means that the method of *arriving* at scientific statements is the inductive inference from observational premises. In the second instance, it means that induction is the unique method of *confirmation* of scientific statements, regardless of how we arrived at them.

Such an understanding of the relationship between facts and theories was typical not only for those simple-minded versions of empiricist philosophy which recommended an uncritical collection of facts and which brought about the various 'natural histories', so characteristic for the beginning of modern science: this understanding may be found also in those versions of empiricism which, being aware of various '*idols of the mind'* deforming our empirical knowledge, recommended certain methods for avoiding these deformations. According to some authors, such a method consisted in a more sophisticated inductive inference (Bacon's induction by elimination, Mill's canons), while according to others it consisted in the possibility of mathematical formulation of relations between facts.

Furthermore, this understanding can be found not only in the programmatic struggle against hypotheses and theories which go beyond the realm of established facts, but also in a more sophisticated opinion according to which all theoretical knowledge is but an epiphenomenon, and should be considered as an economic, convenient means of accounting for facts. Theories are validated when they can be totally reduced to observational statements. This reduction has sometimes been understood as such a 'translation' of theoretical into empirical statements that the finite set of empirical statements exhausts the meaning of the theoretical statement. Sometimes it was understood in a weaker sense, namely in such a way that all the theoretical terms (i.e., the terms which do not denote the directly observable objects, qualities or relations) should be connected with empirical terms by means of some kind of coordination definition or semantic rules. Regardless, however, of the way in which the reduction is understood, this conception means that theoretical statements which cannot be reduced to empirical ones cannot be considered as scientific because they are not susceptible to empirical control. Theoretical knowledge which cannot be reduced to observational statements is empty of any content and without any cognitive value; it is simply a crack through which metaphysics sneaks into science. It might be worth noting that the criteria of demarcation, which we discussed in Chapter II, are meant to delimit just such a set of statements so that any one of the set is either observational or can be reduced to observational statements. The 19th century positivistic attack against the partisans of atomism and of the corpuscular interpretation of the laws of thermodynamics is but one of the more spectacular examples of such an understanding of empiricism and of the empirical method.

The very concept of scientific method has been connected with the belief that it can guarantee the knowing subject an independence from any particular (either individual or collective) biases, and secure the possibility of a fully objective description of reality in the theories he constructs. This belief was the *credo* for the adherents of the version of empiricism we are discussing.

This same belief concerning the invariance of once established facts and the one-sided relationship between theoretical and observational statements can be found in the contemporary attempts to reconstruct the purely empirical language of science. The statements formulated in that language (containing only observational and logical terms) are to provide meaning to all scientific theories. Only those theoretical statements which satisfy some logical conditions concerning their relationships with the statements of the empirical language are to be considered as scientific. This same belief is to be found underlying that theory of scientific explanation which assumes that in the transition from one scientific theory to another, the meanings of observational statements which are to be explained by both of them does not change. This is one of the possible interpretations of the so-called principle of correspondence between theories which we will discuss later.

All those theories of knowledge which assume a one-sided relationship between facts and theories, or between theoretical and observational statements, we label as radical empiricism.

III

If scientific theories are to be empirically confirmable or falsifiable, they must in some manner be susceptible to the verdict of experiment – and this is why they have to be connected in some way with observational statements. Therefore, the first task for a methodology which attempts to reconstruct the empirical basis of science is to single out a set of statements which could serve as a frame of reference for the evaluation of theoretical statements.

According to inductivism, the empirical basis of science is constituted by statements which we accept on the grounds of sense-data. These statements constitute the foundation for justifying all other statements which are not observational and which are usually named *theoretical*.

The empirical basis of science, as opposed to theoretical statements, is to consist of (at least according to some inductivist interpretations) unquestionable statements. These are, of course, not those statements which someone has accepted on the grounds of some sense-data, but those which, just due to their observational character, can be accepted without any need of further justification. (Let us notice that we are disregarding here the problem of understanding these statements either according to phenomenalism, as accounts of the sensual experience of a knowing subject, or according to physicalism, as reports concerning the physical states of the universe.) Hence, it must be specified first, what are the characteristics of these statements, i.e., what kind of sentences we can consider as belonging to the empirical basis of science, and second, how all the other statements are to be connected with this empirical basis.

The solution of these two questions would be equivalent to the logical reconstruction of the structure of scientific knowledge and, as we know, most of the efforts of radical empiricists and especially of logical empiricists were directed towards this task.

In order to solve the first problem, it is necessary to single out or to construct such a language which would contain, aside from logical terms, only observational ones denoting the objects, qualities and relations provided by sense-experience. Only the sentences formulated in this language could be accepted as true or false exclusively on the basis of experience, without recourse to any assumptions or theories.

The solution to the second problem, which will not concern us here at length, would require the formulation of some semantic rules which should relate non-observational terms, which do not belong to the empirical language, with observational ones. Without these rules, sentences which contain theoretical terms would not have any empirical meaning, would not fulfill the criterion of demarcation and it would be impossible to state their relationship to observational statements which have been accepted on the basis of experience. Thus, to give one example, a statement about changes in the energetic state of atoms can be justified only when it is known what kind of directly observable phenomena corresponds to these changes, that is, which observational statements can constitute a basis for the acceptance or rejection of that theoretical claim.

The program of the logical reconstruction of knowledge thus formulated has met with a basic criticism questioning the possibility of isolating or constructing a purely empirical language of science and, by the same token, of isolating an empirical basis which could fulfill the role of an unquestionable foundation of knowledge. The most developed form of this criticism may be found in the works of Popper, whose position has been termed as *falsificationism*, or in contrast to inductivism, as deductivism. We will presently devote some space to this criticism, as it not only will demonstrate the weak points of the radical empiricist theory of knowledge but will also help to observe some of the troubles of the conception of the logical reconstruction of the process of the growth of knowledge which we have discussed in the previous chapter.

Popper puts forward two basic arguments against the program of the logical reconstruction of knowledge. The first concerns the possibility of justifying statements on the basis of experience (*note the justification – not the acceptance*), the second in turn, the possibility of isolating a purely observational language.

The acceptance of a sentence on the basis of experience is not the same thing as its justification. It is one thing to inquire in what manner one has reached a particular conclusion, and quite a different manner in what manner this conclusion can be justified. This is the distinction between the context of discovery and the context of justification, between the questions '*quid facti?*' and '*quid juris?*'. By equating the acceptance of a statement with its justification, the inductivist is committing, according to Popper, the cardinal methodological mistake of psychologism.[1] In order to justify a sentence one can only refer to other sentences, as justification is a logical and not a psychological process. This procedure is, of course, without end. Therefore, no sentences, neither theoretical nor empirical, can constitute an unquestionable, fully justified basis on which our knowledge is built. Despite the conviction of certain inductivists, the empirical basis of science cannot consist of justified, unquestionable statements. Therefore, any scientific statement can be accepted only provisionally. "The game of science is, in principle, without end. He who decides one day that scientific statements do not call for any further test, and that they can be regarded as finally verified, retires from the game."[2]

Obviously, the process of justification cannot be extended indefinitely, we must cut it off at some point. This implies that the problem of which sentences we accept as not requiring, for the present, any further justification, is solved on the basis of a decision by convention. This decision is never final. At any moment we may retract it and demand further justification of an observational statement.[3] In practice, the decision, whether a statement is to be accepted as directly testable or not, is decided by agreement of competent investigators. "And if they do not agree, they will simply continue with the tests, or else start them all over again."[4] Therefore, the point of view advanced by Popper and opposed to the radical empiricism, is a variant of conventionalism.[5]

In criticizing this Popperian opinion, J. Kotarbińska writes:

> The intentions which stand behind this kind of conventionalism are not quite clear. It is directed against psychologism, but against what variant of psychologism? Does the psychologism, which is here denounced, consist in the belief that the choice of the appropriate basic sentences depend upon their observational nature, that it is required that they refer not simply to observable phenomena, but to phenomena that have actually been observed by someone? Or, does it consist in the fact that the observational character of basic sentences is considered as a criterion of their validity, as a full guarantee of their truth? Popper's critical arguments are directed almost exclusively against this second interpretation of psychologism, which is much stronger than the first and which, we may add, is not shared by all the inductivists. Popper's own proposals, however, which he presents as results of this argumentation, seem to bear witness that his tendencies are more radical. They seem to be opposed to any postulate of choosing the basic statements in respect to whatever psychological criteria. This seems to reveal Popper's genuine intentions.[6]

Now, if my understanding of Kotarbińska's argumentation is correct, she states that (a) the inductivist also does not have to consider the observational statements as unquestionable and unretractable, and (b) if science is to be science and not a pure fantasy, then it must be based on some observational statements. The negation of this last thesis by Popper is seen as a result of his overly radical anti-psychologism.

I would be ready to agree with the first part of the argument; undoubtedly, an inductivist does not have to treat observational statements as unquestionable, unless, of course, their unquestionability would be treated *ex definitione* as a characteristic of observational statements.[7] Then, however, the evaluation of the degree of confirmation of theoretical statements would be doubly complicated. The uncertainty of confirmation (inductive inference) is further complicated by the uncertainty of the premises of the inference. (This is not, of course, a remark disqualifying the argument.)

However, the second part of Kotarbińska's argument raises certain doubts. She analyses *junctim* two different Popper theses, namely (1) that the empirical basis of science does not consist of purely observational statements and (2) that observational statements are accepted by convention. She interprets both of them as the results of Popper's radical anti-psychologism, as "the result of treating methodology as one of the branches of logic in the narrow sense of the word, as a discipline which is interested only in logical relations between sentences."[8]

Now, let us remark first of all what is in this context a minor point, that this approach to methodology (which as it should be evident from my previous arguments I do not share) is common also to certain inductivists, certainly to those criticized by Popper. He criticizes them not because they defend psychologism, but because they are unable to overcome it consistently.

What is more important. I believe that the Popperian thesis stating that the empirical basis of science does not consist of purely observational sentences – as Popper says quite explicitly – has nothing in common with his otherwise overly radical anti-psychologism or, as I would prefer to term it, with his logicism. Popper claims, and this is his strongest argument against the inductivists, that purely observational sentences cannot constitute the empirical basis of science *not because they are psychological, but for the simple reason that they do not exist at all.* If by observational sentences we mean such sentences which can be accepted (either ultimately or provisionally) *without referring to any assumptions or theories, then there are no such sentences.*

IV

The majority of statements which are considered as reports of experimental data are established by means of measuring instruments. Now, we must stress that since scientific instruments are always constructed on the basis of some theory, therefore the statements of that theory determine the meaning of observational statements reporting the results of experiments performed by means of the instrument. By the same token, these theoretical statements have their impact upon our accepting or refuting observational statements, and the experimental reports cannot be considered as purely empirical. When Galileo claimed that by means of the telescope he had constructed, he was able to discover mountains on the moon and spots on the sun, and that these observations could not be reconciled with Aristotelian cosmology, the controversy centered not only on the cosmological theories, but also on the optical theory of the instrument, which was seriously questioned by Galileo's opponents.[9]

Furthermore, the acceptance or rejection of an observational statement depends not only on the acceptance of the theory on the grounds of which

the instrument is constructed, but also on opinions concerning the precision and sensitivity of the instrument. Thus, for example, the acceptance
or the rejection of a statement concerning the existence or non-existence
of micro-organisms in the test-tube depends not only on the optical theory
of the microscope we use, but also on the theoretical assumption that if
there are micro-organisms in the test-tube, they would be observable by
means of a microscope. This assumption contains an opinion concerning
the size and the very existence of micro-organisms, as well as an opinion
about the resolving power of the microscope.[10]
    Finally,

as we are aware of the fact that the very act of investigation (for example – of measurement) often disturbs the object under observation (the measured magnitude), therefore,
the distinction between situations when this disturbance can be neglected as irrelevant,
from situations when it should be accounted for, is impossible without accepting some
theory according to which we interpret the results of observation or measurement as
reliable, 'objective', or as necessarily inexact.[11]

The arguments presented above should suffice to accept the fact that no
observational statement accounting for results of an experiment in which
a measuring instrument is used can be accepted without recourse to some
theory or assumptions. Therefore, no such a statement can satisfy the
criteria which radical empiricism imposes on purely empirical statements.
    There remains the possibility of considering as purely empirical those
statements which account for direct observations performed without the
use of any instruments. It would indeed be the case if the results of such
observations could be treated as 'bare empirical facts'.
    Approaching this problem we are faced with the time-honored question
concerning the subjective conditioning of the scope and content of the
sense-data.
    We will neglect here the problem of the individual conditioning which
may result from the dysfunction of someone's cognitive apparatus, as this
is not relevant to our problem. The principle of intersubjective control of
experimental reports, which demands the acceptance only of those statements which may be tested by any appropriately trained observer, robs
this problem of any deeper methodological value, at least in the case of
statements pertaining to the outside world. (We deliberately omit here the
problem of empirical data which are available only by introspection and,
therefore, not susceptible to intersubjective control.)

When we seek to inquire of the subjective conditioning of empirical knowledge, we are in reality asking a different question. We are interested namely in what is the impact of a normally (i.e., in the limits of the human generic norm) functioning cognitive apparatus upon the content and scope of our empirical knowledge. We want to know whether and to what extent this knowledge describes the external world as it is, independently of any act of cognition, and to what extent it depends upon the knowing subject. In fact, it is a question concerning the determination of the content and the scope of our empirical knowledge, not by the individual features of the knowing subject, but by the characteristics common to every human being, and transmitted from one generation to another. The question is important for dependent on the answers to it (at least in the frameworks of some versions of empiricism) is the solution of the controversy concerning the possibility of achieving objective knowledge. As we know, this problem has found different solutions within empiricist philosophy. So-called naive realism assumed a full conformity between sense-data and the features of perceived objects; critical realism either distinguishes between primary and secondary qualities, assigning objectivity exclusively to the former, or treats all the qualities as dispositional and assumes no more than a certain coordination between the objective features and the sense-data; finally, epistemological idealism considers, as Berkeley did, the experienced objects as part of the complex of our sense data, depriving them of any independent existence.

For reasons which will become clear below, it is not this problem either which is central to our discussion. For the time being, we may assume that no matter which of the proposed solutions of the problem of relationship between the features of external objects and the sense-data is to be accepted, it is just this very knowledge, stigmatized more or less by human cognitive possibilities and limitations, which constitutes the domain of empirical facts we might deal with. These are those 'bare facts' which an empiricist is prone to take as something final and decisive. Now, the question we want to ask might be formulated as follows: Is it true that we ever deal with facts whose content is determined exclusively by objective features of the object and by the generations-invariant unchangeable genetically determined functioning of the human cognitive apparatus? Is this content really independent of the changing, socially determined and inherited intellectual equipment of the knowing subject?

Few would question today that the scope and content of empirical facts in commonsense knowledge also depends on a number of factors which are deteimined not genetically but socially, i.e., on cultural factors. There can be no doubt that the development and the propagation of scientific knowledge is one of these factors. However, the conviction that scientific method can free us from this type of cultural determination, that scientific facts stated according to this method may be free of any 'external' bias and constitute the 'bare facts' in the indicated sense of the term, this conviction is still alive today. It is hard to deny such an obvious fact that the application of scientific method caused and still causes the objectivization and universalization of our knowledge, and that the progress of science would be otherwise impossible. It may be doubted, however, whether this objectivization of commonsense knowledge achieved by the application of the scientific method allows us to treat scientific facts as 'bare empirical data'.

We have every reason to believe that the capacity of the knowing subject for conceptual thinking, for the use of language, for communication with other subjects, and for the production and utilization of things is not without effect on the manner of perceiving natural phenomena at every stage of cognition. These capabilities are realized, however, in different situations and conditions. Therefore, we may, and have to, ask not only how the content and the scope of our empirical knowledge are affected by the circumstance that raw sense-data are always presented by means of *some* conceptual apparatus, expressed in *some* language and, accordingly, understood in some particular manner; we also may, and have to, ask how our understanding of natural phenomena provided by experimental evidence is affected by the circumstances that the knowing subjects are endowed with *this* conceptual apparatus they actually use, that they express their judgments using *this* particular language and not another. The questions we ask, the nature of the answer we receive and how we understand them, depend upon all these factors; by the same token they determine the scientific facts we are dealing with.

Therefore, even when we do not use any measuring instrument in the course of an experiment, the result is interpreted, and the observational statement is accepted or refuted, on the grounds of different assumptions and theories concerning the structure of the world and the mechanism of our cognitive functions. Just as we accept or refute statements stating the readings of a scientific instrument not only on the basis of observation, but also on the grounds of theories pertaining to the functioning of the instrument, so statements reporting on direct observations and expressed in a given conceptual apparatus are accepted or refuted on the grounds of certain theoretical assumptions concerning the functioning of our cognitive apparatus, concerning the reliability of information it provides, etc.[12] This is obvious in the case when we question, for example, the reliability of someone's observational reports by taking into account a particular, 'abnormal' state in which the observer found himself in the process of experiment or observation. If we accept such a statement as a reliable one, we do this assuming that no disturbing circumstances were present. Moreover, we accept some theoretical opinions compelling us to accept observational statements accounting for sense perceptions, if they were stated under 'normal conditions'. Thus, also in the case of observation statements accounting for the results of direct observation, the situation is not different from that when we use instruments. At most in such a case the theoretical assumptions we accept are less evident and are accepted without stating them explicitly.

In a paper written by Ajdukiewicz in 1934, we read:

> The unique difference that exists between observational and theoretical (interpretative) statements is that the first are evaluated as true or false by means of languages we are endowed with without our participation, while the latter are evaluated by means of languages we have consciously constructed by ourselves. It is why the semantic rules by means of which we evaluate observational statements seem to be unquestionable, whereas the convention needed for the evaluation of theoretical (interpretative) statements, as introduced by an act of our will, seems to be revocable by our decision.... *I do not see any difference between observation and theoretical statements. I believe that mere experimental data do not compel us to accept either one or the other.*[13]

A few years later, the author of these words abandoned the point of view of radical conventionalism[14] he defended in the paper quoted above. However, he did not abandon the opinion we have referred to. In a paper published in 1953, he wrote:

> I would not like to be misinterpreted. The abandonment [of the position of the radical conventionalism] does not signify the rejection of the weak thesis which was intended mostly to stress the fact that in order to construct any sentence it is necessary to use some language, and in order to construct any judgment it is necessary to use some conceptual apparatus.[15]

Therefore, even the most 'bare fact' is clothed in some theoretical garb, if only because (and this is not the only reason) there is no experience

without some theory of experience which determines how experimental
results are to be interpreted, there are no judgments without some
conceptual apparatus by means of which we express their content, there
are no statements without some language by means of which we formulate
them. If this cloak appears to be transparent, it is rather a result of
neglecting the epistemological reflection than a consequence of its
genuine 'transparence'.

V

As should be clear from the foregoing argument, the position of inductivism is not maintainable regardless of whether observational statements
are considered as questionable or not. It is simply not true that the
relationship between the empirical basis and theories is one-sided. Just
as the observational statements constitute the basis for accepting or
refuting a theory, so the previously accepted theories are involved in the
formulation, acceptance or refutation of observational statements. And
if there are no purely empirical statements, then it is impossible to maintain the position that scientific theories are accepted or refuted on the
basis of their inductive confirmation only. I believe that in this respect
the Popperian criticism of inductivism is cogent.

If we reject the radical empiricist conception of a purely empirical
basis of science, if we agree that the empirical foundations of science are at
least partially theory-laden, that scientific theories are neither an extract
from experience nor only its convenient, economic description, but a
manner of viewing the universe which codetermines the interpretation of
any experience, then the question of the empirical determination of the
content of knowledge must be replaced by the problem of mutual relationships between facts and theories.

Furthermore, the fact that radical empiricism has overcome Kantian
apriorism does not mean that it has provided a solution for the problem
of the all-encompassing character of theoretical assumptions, the problem
in which Kant was so keenly interested. Radical empiricism tried to
cancel this problem rather than to solve it. It did not acknowledge that it
is not only the discoveries of new facts that constitutes the driving power
of the growth of science, but also the appearance of new ideas which
compel us to reinterpret the known facts. These lose, of course, their
alleged invariance.

In accepting the empiricist solution to the problem of the mediation of the senses in every act of acquisition and control of knowledge, we are by no means forced to state that the knowing subject confronts nature directly. Rather he confronts it armed with knowledge inherited from his predecessors, with convictions about the structure of the universe and the place the object under study occupies in that universe. He is endowed with convictions as to the human cognitive capacities, with a tradition of practicing science, i.e., with a scientific method, beliefs concerning the goals of science and its social role. No scientific method, even a best one, can neutralize these convictions and beliefs, if only for the reason that it is itself their product.

Though all this endowment is not *a priori* to the world of human experience, nevertheless it is *a priori* in respect to the experience of a particular knowing subject who undertakes a scientific problem, and it has a definite impact upon how he will solve it. This is, of course, a specific kind of '*a priori*'. It is imposed upon the knowing subject neither by the 'eternal laws of thinking', nor by the categories of pure reason, which according to Kant were the necessary conditions of intersubjective cognition. It is imposed rather by a concrete but changeable socio-cultural situation in which the individual happens to live and work, which he encounters and accepts partially unconsciously, but which can be and often is an object of intellectual criticism and analysis, due to which it may change. And it is epistemology which performs this function of analysis and criticism. But this function can be accomplished only by an epistemology which does not assume the autonomous character of experience and does not absolutise its results.

In order to avoid any terminological misunderstandings, let us try to express this thesis in a somewhat different manner. The thesis that experience is prior to any theory is ambiguous. It is most probably true when it signifies that any theoretical concept has ultimately some empirical sources. This is equally as true of myths as of the theoretical constructs of science. This thesis is false, however, when it is understood in the sense that experience is prior to theoretical knowledge in each and every concrete act of cognition. In this case quite the opposite is true. An experiment is undertaken on the grounds of some theoretical knowledge[16] which is prior to, and usually, independent from it and in this sense, *a priori*. It seems important to stress that this *a priori* knowledge consists not only of theories accepted in a given scientific discipline, but of a much wider range of theoretical beliefs.

The rejection of the thesis that the empirical basis of science consists of purely empirical statements accepted (finally or provisionally) on the grounds of experimental results, of statements which are not theory-laden, abolishes the absolute distinction between the theoretical and empirical languages. The distinction between 'empirical' and 'theoretical' terms may be only a relative one. It is relative historically as well as in respect to various linguistic systems, i.e., in respect to different theories. A scientist who undertakes the study of a particular problem, for example of a biological one, and who uses various scientific instruments constructed on the grounds of different physical theories, is quite aware of the fact that together with the equipment he uses he accepts also these theories. In spite of this fact, however, he will treat the statements he will formulate by means of these instruments as observational. The observational language is, for him, something already present and historically given by the development of science and common knowledge. He will not be concerned with the fact that certain terms he uses in formulating his observational statements are treated as theoretical by physicists and are in need of further empirical interpretation. If, however, on the basis of his experiments, in which dozens of physical, chemical and other theories were intertwined, our biologist would attempt a new theoretical interpretation of the facts he has stated, and in order to do this would be forced to introduce some new concepts, or to reinterpret the old, he would feel obliged to give them some observational meaning. He will then specify the relationships between these new terms (or the expressions containing them) and the terms which are considered in the language of his discipline as observational and in no need of further interpretation.

NOTES

[1] K. R. Popper, *L.S.D.* p. 95, 102.
[2] *Ibid.*, pp. 53–54.
[3] *Ibid.*, pp. 104–105.
[4] *Ibid.*, p. 104.
[5] Cf. the next chapter.
[6] J. Kotarbińska, 'Kontrowersja: dedukcjonizm-indukcjonizm' (The Controversy: Deductivism-Inductivism), in *Logiczna Teoria Nauki* (ed. by T. Pawłowski) Warszawa 1966, pp. 324–325.

7 Criticizing the inductivist point of view, J. Giedymin presents their conception of observational statements as follows: "The meaning, and by the same token the condition, for the truth of observational statements are completely determined by perception. A refutation of a statement despite the sense-data or despite an observational situation coordinated to the statement by a semantic rule leads to a violation of its meaning." ['O teoretycznym sensie tzw. zdań obserwacyjnych' (The theoretical sense of so-called observation statements), in *Teoria i Doświadczenie* (Theory and Experience) ed. by H. Eilstein and M. Przełęcki, Warsaw 1969, p. 92.]

8 J. Kotarbińska, *op. cit.*, p. 325.

9 Cf. B. Kuznetzov, *Galileo*, Moscow 1964, Chapter IV.

10 Cf. J. Giedymin, *op. cit.*, p. 107. This paper gives an excellent analysis of a number of historical cases. The analysis presented here shows that all the so-called observational statements, on the ground of which scientific theories were accepted or refuted, were in fact theoretical in character, i.e., that the theories were not accepted or refuted *only* on the grounds of experiments.

11 *Ibid.*, pp. 107–108.

12 Cf. K. Pomian, *Przeszłość jako przedmiot wiary* (The Past as an Object of Faith) Warsaw 1968, especially Chapter I which provides an excellent analysis of the impact of epistemological assumptions upon the acceptance of historical reports.

13 K. Ajdukiewicz, 'Das Weltbild und die Begriffsapparatur', *Erkenntnis* 4 (1934), 274 [my italics – S.A.].

14 Cf. K. Ajdukiewicz 'W sprawie artykutu Prof. A. Schaffa o moich poglądach filozoficznych' (Reply to Professor A. Schaff's paper on my philosophical opinions) in K. Ajdukiewicz, *Język i poznanie*, Vol. I, Warsaw 1965, p. 176.

15 *Ibid.*, p. 181 (footnote).

16 The thesis that in testing a theory we confront it with theory-laden statements should be clearly differentiated from the stronger thesis that this theoretical basis comes from the tested theory itself. The first thesis leads to the conclusion that no test (either verification or falsification) is conclusive – i.e. that the so-called asymmetry between verification and falsification is apparent. The second thesis would imply that no falsification is possible at all. It seems that the Kuhnian conception of paradigms as self-confirming entities comes from this stronger thesis, while the arguments he provides confirm only the weaker. [For details see Chapter VI.] I am indebted for this remark to Dudley Shapere.

CHAPTER V

FACTS AND THEORIES: CONVENTIONALISM

I

In the last paragraphs of the previous chapter, we explained why we believe that Popper's criticism of the radical empiricist solution of the problem of relation between facts and theories is correct. The Popperian point of view in this matter is, as we have noticed, a variant of conventionalism. Despite a number of interesting details in Popper's proposals, his solution of this problem is not at all surprising: the criticism of the concept of 'bare empirical facts' is to be found within all conventionalist doctrines, in the writings of Poincaré [1] and Duhem, [2] as well as in the works of Adjukiewicz stemming from the period when he still held this position. We must remember, however, that Adjukiewicz did not abandon this criticism even when he gave up the position of conventionalism. As I do not believe either that the alternative to radical empiricism of some kind of conventionalism exhausts all the possibilities for solving the problem, I will now discuss it in more detail. With this intent I will first present Popper's conventionalist solution of the problem of the relation between facts and theories and then, in the concluding paragraphs of this chapter, discuss in general terms the controversy between empiricism and conventionalism.

Popper's conception of the growth of knowledge, and the strategy of scientific procedures he proposed were rightly called the strategy of permanent revolution. This strategy assumes that acquired knowledge is never perfect and that the best way to improve it consists in submitting it permanently to the most severe tests. Theories which cannot withstand these tests should be rejected and replaced by better ones. This strategy, however, can be applied only if the claims of science are falsifiable, i.e., if it is possible to attempt to disprove them. Statements which, for whatever reason, cannot be submitted to this kind of testing and which, therefore, may be maintained regardless of the results of experiment, do not fulfill the criterion of demarcation, are unfalsifiable, and consequently cannot be treated as scientific. This holds both for theories which do not have any empirical content and, therefore, can be neither confirmed nor falsified, as well as for theories which have empirical content but are endowed with some kind of self-defense mechanism due to which they are compatible with any possible result of experiments and can never be falsified by empirical evidence. As examples of the latter kind of theories, Popper has usually quoted psychoanalysis and Marxism. "The problem which troubled me at that time [in 1919] was neither 'When is a theory true?', nor 'When is a theory acceptable'. My problem was different. *I wished to distinguish between science and pseudo-science.*"[3] The criterion of demarcation we mentioned above, namely falsifiability, was believed to provide the solution to this problem. It is based on the presumption that the infallibility of a statement is not its merit but a fault, that "the virtue does not consist in avoiding mistakes but in their inexorable elimination".

If science is to be a rational activity, we must be able to state in respect to any scientific statement what are the experimental data in the face of which we would be ready to abandon it. If we were unlikely to specify a possible experimental result which, if obtained, would compel us to refute the statement, it would mean that the reasons by which we have accepted it are not rational and that the statement itself is metaphysical and not susceptible to scientific procedures. If scientific activity is to secure the growth of knowledge, it must proceed by permanent attempts at disproving accepted opinions and abandoning of those which have been falsified. Anyone who exclusively seeks evidence confirming the opinions he already holds accepts them as a believer and not as a scientist. Confirming evidence may always be found, but no confirmation is ever conclusive. It does not advance our knowledge, at best it consolidates positions already achieved. Falsification, on the other hand, compels us to look for new, better theories. Assuming that any theory must finally be found limited and in need of some revisions and conjectures, the Popperian strategy prefers the most risky, improbable hypothesis as the most susceptible to falsification and, at the same time, recommends the most critical attitude towards them.

The concept of a risky hypothesis needs a more detailed explanation. Following Popper we will single out a class of singular[4] existential statements of the form 'Such a phenomenon happens in the space-region k', which he calls *basic statements*. These statements satisfy the following condition: (a) they cannot be deduced from universal statements without additional assumptions, (b) they can contradict universal statements or their consequences.[5]

Any theory (system of universal statements) splits the class of basic statements into two subsets: the first consists of statements contradicting the theory, that is of those which would falsify it if they were accepted on the basis of experience. These constitute the potential falsifiers of the theory. The second subset contains basic statements which do not contradict the theory. These are either in agreement with it or are simply neutral.[6]

The set of potential falsifiers is the more numerous, the more risky the hypothesis is. If the relevant class of potential falsifiers is empty, we risk nothing in advancing the hypothesis. No experiment, regardless of its results, can falsify it. Such an experiment is of no value from the point of view of the development of knowledge. A theory which allows only for such experiments does not fulfill the criterion of demarcation; it is metaphysical and unscientific.

For the sake of illustration let us compare science with the game of roulette. The equivalent of the result of an experiment will be, for us, the number on which the ball stops. The equivalent of the hypothesis being tested[7] is the statement: the ball will stop at the number $n (0 \leqslant n \leqslant 36)$ which belongs to a subset p of the possible results (we may play for one number or for many numbers at the same time). The basic statements will have the form: 'On this wheel the number n came out.' Our hypothesis divides the set of basic statements into two subsets: the first consisting of statements being in agreement with the hypothesis, the second of statements contradicting it. Depending on which basic statement we accept on the grounds of an experiment, our hypothesis will prove to be either true or false. Now the postulate of advancing the most risky hypothesis is equivalent with the following strategy of playing roulette: "Bet upon the less numerous subset of the set of possible results". Or putting it more simply: "Bet rather upon a single number than upon a dozen, rather upon a dozen than upon the even or the odd numbers, but never bet upon all the possible outcomes at the same time because in this case it is sure you will guess the result, but you will win nothing". The more risky the hypothesis, the greater the prize in the case of success. In scientific activity, the success consists in advancing a hypothesis which can withstand the test, and the prize is new information.

The same idea may be expressed in still another manner: the more numerous is the subset of potential falsifiers, i.e., the more the hypothesis prohibits, the more risky it is. Accordingly, Popper sees scientific laws as prohibitions, as statements specifying not 'what is happening', but 'what must not happen'. The occurrence of a 'prohibited phenomenon' is a falsification of the law which prohibits it.

The measure of risk of a hypothesis (in respect to previously acquired knowledge) is estimated by Popper by means of his own interpretation of the calculus of probability. Accordingly to this interpretation, he states that the more improbable the theory is, the more valuable it is. While this may sound shocking, it must be noted that 'a highly probable hypothesis' is not a hypothesis strongly confirmed by performed experiments but one which is less exposed to possible future falsification. And, by analogy, an improbable hypothesis is not one that received weak confirmation, but one which may be submitted to a large number of various tests, every one of which might eventually falsify it. On the grounds of this interpretation, a hypothesis with probability $p = 1$ is of no cognitive value: any future experiment can only confirm it, or no test can falsify it. The first possibility occurs when we are dealing with a theory endowed with a self-defense mechanism, the second when we are dealing either with such a theory or with a theory of no empirical content at all. In this last case both confirmation and falsification are impossible.

This conception of estimating the probability of hypotheses is opposed to Carnap's, according to which the probability of a hypothesis is the degree of its inductive confirmation, as well as to Reichenbach's which interprets it as a 'weight' of a hypothesis, as its degree of truthfulness. According to Carnap, the degree of probability of a hypothesis is measured by the degree of its confirmation on the grounds of already acquired empirical evidence.[8] Therefore, the most valuable are those hypotheses which are strongly confirmed, highly probable. (A statement with the probability $p = 1$ is a statement which follows deductively from accepted evidence.) It is hard to deny that this usage of the term 'probability' is closer to the usual understanding of the word. This, obviously, is not an argument against Popper's interpretation but rather evidence that the usual understanding grew out of the inductivist concept of knowledge which Popper refutes.[9]

What is not subject to any doubt here is that each of these theories is associated with a totally different conception of scientific method. The Carnapian grows out of the belief that scientific hypotheses are accepted on the force of their inductive confirmation which is measured by logical probability. Popper's conception, on the other hand, is based on the opinion that hypotheses in science are tested deductively. This means that they are tested by deducing their consequences from them and confronting them with basic sentences already accepted on the grounds of experiments, that is by attempts at their falsification. This procedure is often called the 'method of criticism of hypotheses'. In reality, it is a variant of the method of 'trial and error'.

II

Now, (a) if theories in science are evaluated by confronting them with observational statements: (b) if these sentences are not, as we have stated, purely empirical because their acceptance depends partially upon accepted theories and assumptions, and (c) if an experiment cannot (in the logical sense of the term) justify a basic statement but can only serve as a motive for its acceptance, then we must face the following questions: first, what decides at which moment we have to stop the process of justification and agree that a basic statement has been corroborated by experiment and second, what is, in fact, verified or falsified by an experiment?

When we accept a basic statement on the grounds of experimental evidence and of more or less consciously accepted assumptions, we may always go further and ask for the justification of these assumptions. In principle, this process may be continued without end but, as a matter of fact, we stop it at some point. Popper's answer to this problem is the following: the choice of basic statements is a matter of convention. On the one hand, we need a convention according to which we accept that the given sentence is a basic one, i.e., empirically testable. On the other, we need a convention on the force of which we temporarily stop the process of justification at this testable statement.

By the same token it appears that the structure of an experiment is much more complicated than the adherents of radical empiricism believe it to be. What is confronted with the obtained result is not simply the theory under test, but the theory in conjunction with various theoretical assumptions (background knowledge) which participate in the interpretation of the obtained result and determine the decision whether to accept or refute the observational statement. The principle of falsification is based on a classical *modus tollens*:

$$[(T \to P)\cdot(\sim P)] \to (\sim T)$$

As, however, it is not the theory alone which is confronted with experiment, but the theory in conjunction with background knowledge, we are dealing in fact with the following situation:

$$[(T\cdot Z \to P)\cdot(\sim P)] \to \sim(T\cdot Z)$$

Thus the result of the experiment is not conclusive. We do not know what is actually falsified, the theory (T), which we are testing, or the background assumption (Z). We know that something is wrong in our theoretical knowledge, that we cannot accept both (T) and (Z), but we do not know which of them is wrong. We are aware that something must be corrected, but we do not know what is to be modified. The assumption that the experiment falsifies the tested theory is a result of a convention which separates the theory in question from its background knowledge. It is a convention according to which we agree to consider all of the background knowledge as unquestionable and decide to throw the theory under test at the mercy of experiment. Thus it is apparent that no experiment can be, nor is, an *experimentum crucis sui generis*. If we decide, however, to accept it as such, we do this with the force of a convention. We could just as well maintain the theory and introduce some modifications into the background knowledge. *Eo ipso*, any theory can be saved from falsification, either by introducing new assumptions or by modification of the already accepted ones. Such a falsification is not any more conclusive than confirmation, and the thesis about the asymmetry of falsification and confirmation seems to be questionable. Just as confirmation does not prove that the theory (T) is true, so falsification does not prove its falsehood. What results from this situation is, according to Popper, a methodological postulate prohibiting the saving of a theory from falsification through the introduction of some *ad hoc* hypotheses or assumptions.

It is clear then that Popper gives a conventionalist answer to both of
the questions formulated above. Methodological conventions are to
decide which sentences are to be accepted as basic and in no need of
further justification, as well as which part of our knowledge involved in
the experiment is to be assumed as wrong when we find a negative result
from an experiment.[10]

It is just this conventionalist solution, and not the refutation of the
conception of a purely empirical basis of science, that is, I believe, the
consequence of Popper's overly radical logicism, of his refusal to go
beyond the realm of logic in the philosophy of science. It is just this
problem – why some statements, either theoretical or basic, are accepted
in science in such a way that the majority of scientists consider them as
valid in spite of the fact that they are not definitely justified and, therefore, may be questioned or defended by means of additional assumptions –
which cannot be solved within the framework of the logic of scientific
discovery. We are faced with an alternative: either we have to go beyond
the context of justification, or we accept some kind of conventionalist
solution. We may even phrase it more thoroughly: the conventionalist
solution provides no explanation at all. It is simply a consequence of
restricting the analysis of the growth of knowledge to the context of
justification and furthermore of the prohibition from asking further
questions which cannot be of strictly logical character, i.e., pertain only
to relationships between statements, but must take into account historical,
sociological and, quite possibly, pragmatic factors of the growth of
knowledge.

It is just at this very point that actual scientific practice does not
conform with the normative pattern of the logic of scientific discovery.
And this situation is not a result of accidental or incidental reasons, but
of factors which are fundamental. The development of knowledge is not
a process which occurs in the 'third world', and, therefore, its course is
and must be determined not only by logical factors.

If I remarked previously that the breakthrough which Popper made in
the traditional practice of philosophy of science paved the way for
questions whose answers do not fit into the framework of the Popperian
program of the logical reconstruction of the process of growth of knowledge, it is this very point I had in mind. The well-justified refutation of
assumptions shared by the radical empiricists compels us either to go

beyond methodological problems in the philosophical reflection on science and especially upon its development, or to seek a solution in some version of conventionalism. Lakatos, who *nota bene* accepts Popper's basic assumptions, is well aware of this situation when he writes:

> One alternative is to abandon efforts to give a rational explanation of the success of science. Scientific method (or 'logic of discovery') conceived as the discipline of rational appraisal of scientific theories – and of criteria of *progress* – vanishes. We may, of course, still try to explain *changes* in 'paradigms' in terms of social psychology. This is Polanyi's and Kuhn's way. The other alternative is to try at least to *reduce* the conventional element in falsificationism (we cannot possibly eliminate it)....[11]

This appraisal of the situation seems to me to be quite to the point. We must either go beyond the logic of scientific discovery or accept a version of conventionalism. This does not, of course, imply that I agree with Lakatos that the expansion of the philosophy of science beyond the study of the context of justification is equivalent to resignation from rational explanation of the successes of science. This would only be so if criteria of rationality in science were invariable and were of themselves without need of explanation. The rational explanation of the successes of science depends, as I see it, on the attempt to understand those variable criteria of rationality which have actually guided its development.

The refutation of the conception of a purely empirical basis of science, the recognition that every statement is theoretical in character, undoubtedly constitutes an achievement of falsificationism and makes it the most important position in the trend of philosophy of science which accepts that this discipline is or should be limited to the study of the context of justification. At the same time this position, because of its normative character, is a proposal for a standard for a scientist's intellectual integrity, for his ethics *qua* investigator. But it is this very principle that leads to its difficulties. Popper's diachronic epistemology has at this point opened the door to historical questions, while his conventionalism, as well as the less radical conventionalism of Lakatos, (which we shall discuss in the next chapter), attempts to bar that door.

III

Let us see the difficulties which result from Popper's solution of the problem of the relation between facts and theories.

First of all, we should notice that despite rigorous methodological postulates, his model of the development of knowledge operates with a 'weak' concept of falsification. Falsification, as he interprets it, does not prove the falsehood of the theory under test; there is always a possibility that what is wrong is not the theory but the background knowledge. Therefore, we might say that an experiment brings a theory under suspicion rather than falsifies it, that its verdict may be only circumstantial. Such a falsification does not compel us to abandon the theory unless, of course, we accept by convention that the background knowledge is unquestionable. But in accepting this convention, we agree to bring about the heaviest verdicts on the grounds of circumstantial evidence. Thus, accepting the Popperian requirement of refuting a theory if it is contradicted by a basic statement, we run the risk of abandoning a true theory. This requirement also does not take into consideration the fact, which we mentioned earlier, that new theories, as a rule, are bound by an ocean of anomalies with which they come to grips only in the course of time.

In practice, the scientist is always confronted with a situation in which he must decide whether to refute the theory under test on the basis of circumstantial evidence, or to defend it by introducing new assumptions. This situation arises whenever the evidence from an experiment testifies against a theory. Popper's methodological postulates forbid the defense of a theory in such a situation. The history of science, however, indicates that scientists sometimes choose one solution, sometimes the other, and that either one of them may lead to success or to defeat. It seems that no methodological rule prescribes procedure in such a situation or even serves as a basis for the *ex post* reconstruction of the process of the growth of knowledge.

The fact that stones fall vertically, that buildings and fortifications do not crumble, that wine does not spill from a jar, could all serve equally well as evidence against Copernican theory or against the Aristotelian physics on the grounds of which these facts were considered as being in contradiction with the heliocentric hypothesis. One could deduce from this that the Copernican theory was wrong as well as that something was wrong with Aristotelian physics. One could either state that "As fortifications do not crumble, the earth can move neither around the sun, nor around its axis", or ask "how does it happen that although the earth moves, wine does not spill from a jar, fortifications do not crumble, etc.?"

Which one of these questions was actually asked depended not on the accepted conventions but on certain historical circumstances, including certain 'extra-scientific' convictions of the times. One could well imagine, for example, that if the heliocentric theory had been advanced before the inclusion of Aristotelian cosmology into the commonsense world view, it would be not the astronomers but the physicists who would have found themselves in a difficult situation. It is they who would have had to explain why wine does not spill from a jar in spite of the fact that the earth rotates, and not the astronomers that as it moves, fortresses do not crumble. The *onus probandi* would have fallen on someone else. And in this case it is not Aristotelian physics but the heliocentric theory which would had constituted the 'unquestionable' background knowledge. Even if we assume that the advancement of a scientific discipline must proceed in a certain logical order, this thesis certainly does not concern the whole of our knowledge (*nota bene*, it would be hard to explain what this logical order actually is; at best it could be reconstructed only *ex post*, but not predicted beforehand). And if this is so, then the content and the scope of the background knowledge, as well as the degree of our trust in it, are mostly accidental in respect to the theory which we are testing at the given moment. It seems that just these accidental circumstances rather than the methodological rules of accepting and refuting theories decide whether the scientists refute the theory under test or try to save it through the introduction of modifications and *ad hoc* hypotheses into the background knowledge.

By analogy, the fact that the orbit of a planet did not confirm Newton's theory might have been interpreted to mean that the theory is in error, but could also have led to the hypothesis that there exists some unknown planet which, according to that theory, affects the orbit under observation, that is to a modification of the background knowledge. As we know, Bouvard's trust in Newton's theory led him to advance the hypothesis (in 1821) that the orbit of Uranus was disturbed by some other celestial body and then to the discovery of Neptune by Leverrier (in 1866). But we also know that the irregularity of the orbit of Mercury, despite many years of attempts at explaining it on the basis of Newton's theory, could not be resolved and was explained only by means of the general theory of relativity. Thus we might say, but only *ex post*, i.e., after the problem was solved, that the 'irregularity' of Uranus' orbit did not, while Mercury's orbit did, constitute an *experimentum crucis* in respect to Newton's theory. And the result is that in one case success was achieved by introducing an *ad hoc* hypothesis concerning the number of planets in the solar system and, in the other case, through a modification of the theory itself.

In order to understand how scientists proceed in such situations, it is not enough to refer to known empirical facts. Neither is it enough to appeal to the methodological rules postulated by falsificationism or, more generally, by any logic of scientific discovery. What must be done is the reconstruction of the criteria of rationality accepted, for example, by the partisans and opponents of Copernicanism. It becomes clear then that they tried to embrace the same empirical facts through different conceptions of the 'cosmic order' which they considered as rational and that they considered them as rational for reasons which generally, and especially from the falsificationist point of view, could not be treated as scientific.

In short, history of science does not confirm that the more famous falsifications of theories were accomplished according to the rules recommended by falsificationism. At times scientists were very 'stubborn' and refused to give up a theory even though it was contradicted by certain well known facts. At other times, to the contrary, they abandoned them rapidly. As an example for the first situation we may cite the fact that Newton's mechanics was still sustained for eighty-five years after the discovery of Mercury's perihelion disturbance. The second may be illustrated by Bohr's proposal of a theory of radiation which contradicted Maxwell's well-confirmed theory, or by Galileo's acceptance of Copernicanism. No methodological rules can *a priori* determine the solution in such cases, nor can they serve *ex post* as a basis for the reconstruction of the logical pattern of the process of growth of knowledge.

This, however, does not exhaust the matter. Both the attempts to save a theory by means of introducing *ad hoc* hypotheses or by modifying the background knowledge may end in a fiasco. What does this imply? Does it mean that the theory cannot be rescued or that it was rescued ineptly and further attempts to this effect should be continued? It is obvious that they can be repeated *ad infinitum*. How are we to decide whether the attempt at saving the theory was unsuccessful because of an error committed by the scientist, because of his incompetence, or that maintaining the theory was a collective error? Is there any sense in talking about an error when we are not dealing with a mistake committed by an individual scientist, an error which can always be corrected by his colleagues or even by himself, but with a view which, at the given time, is commonly accepted by almost all competent scientists who in a specific manner see, interpret and understand observed facts and their relation to accepted theories?[12]

Furthermore, is there really a specific moment at which it is possible to state that further attempts at saving the theory are no longer rational, that they result from dogmatism, from an unscientific attitude? The resistance of scientists to new theories, which we are often inclined to judge as a symptom of their dogmatism, is often grounded on strong 'empirical' reasons. (*Nota bene*, resistance against fashionable but erroneous theories is usually treated as a normal phenomenon and seldom reported.) Usually they are able to present a large number of facts or of possible experiments which the new theory cannot handle because they are interpreted in terms of the old theory. If this were not the case, their resistance would be of no importance and could be disregarded. If the partisans of the Copenhagen interpretation of quantum mechanics had seriously to take Einstein's opposition under consideration, it was precisely because Einstein again and again was able to present certain thought-experiments whose explanation in terms of the new theory was neither simple nor obvious and led often to internal inconsistencies.[13] Such a resistance, however, results usually from the inability to view well known facts in a new way, that is in terms of new theoretical concepts. In science, at least, dogmatism seldom consists in denying facts. There are probably few more dangerous sources of error than the inability to grasp what in a given 'fact' is a result of the theoretical interpretation to which we have become accustomed. And only when we abandon a theory does it appear that an object seen previously as a star becomes a planet, that a chemical reaction treated as a decomposition becomes a synthesis, that a movement considered as aiming at a natural place is seen now as an inertial motion, that an inertial motion along a cyclic orbit becomes a rectilinear motion disturbed by universal gravitation and after the successor revolution, a motion along a geodetic in Riemann's space.

Finally, is it possible to differentiate among the methodologically acceptable and unacceptable modifications introduced in order to save a theory? Let us notice that for the sake of saving a theory it is possible to question all the accepted assumptions which constitute the background knowledge: the exactness of the experiment; the theories on the grounds of which the instruments function that are used in the experiment; the ontological assumptions concerning the structure of the universe which affect the way in which the experiments are conducted and the interpretation of their results; the convictions pertaining to the character of cognitive procedures; and even the rules of inference by means of which we state the incompatibility of the tested hypothesis with experimental data.

Of course, it could be said that for the development of science it would be the best if some scientists proceeded in one manner and others in another, which is what is usually the case. But this is not the answer that we expect from a methodology which aspires to present a logical reconstruction of the process of testing, accepting and refuting of scientific theories.

All these questions and problems seem to testify to the fact that the falsificationist model of the growth of knowledge, precisely because it resorts to conventions when it has to deal with issues that lie beyond the sphere of logic, cannot adequately realize the program of the historical reconstruction which it has advanced.

IV

The criticism of falsificationism presented above demands a short digression concerning our interpretation of Popper's views. Imre Lakatos, in his essay *Falsification and the Methodology of Scientific Research Programs*, distinguished between the naive, the dogmatic and the sophisticated versions of falsificationism. Roughly speaking, dogmatic falsificationism presumes the unquestionability of basic statements. There is no doubt that Popper never represented this position. The difference between naive and sophisticated falsifications consists in the following two points: contrary to the naive version, sophisticated falsificationism does not assume that only falsification is an interesting result of experiment; secondly, it states that there can be no falsification as long as a new theory has not been accepted. It means that the acceptance of a basic statement contradicting the theory under test neither falsifies it nor compels us to abandon it.

It is not difficult to see that Popper's position, as we have presented it, conforms rather to the naive than to the sophisticated version of falsificationism. Lakatos, however, says that Popper, and he himself, represent the latter. I believe that my presentation adequately summarizes the basic points of Popper's views as expressed in his *Logic of Scientific Discovery*, in both the original German version and the English translation (of 1956). However, in his book *Conjectures and Refutations*, (from 1963), as well as in *Objective Knowledge* (1972), both of which contain papers written in the last decades, we may find opinions confirming that he, in fact, is willing in certain respects[14] to modify the views expressed in *The Logic of Scientific Discovery*. And some of these modifications, especially in *Conjectures and Refutations*, may perhaps be interpreted in the manner in which Lakatos has done. (The modifications which we find in *Objective Knowledge* are in themselves worthy of a detailed analysis but I believe they do not concern the difference between naive and sophisticated versions of falsificationism. They testify rather to more general changes in Popper's philosophy.)

In *Conjectures and Refutations*, Popper acknowledges that a good theory should not only be a very risky one, but should resist for a given time all attempts at falsification.[15] This means, of course, that the author acknowledges the cognitive value of confirming evidence. He also formulates certain conditions which are to serve as criteria for evaluating whether the new theory is better than its predecessor and should, therefore, replace it.[16] In such a case, however, the evidence which falsifies one theory confirms the other and the whole dispute between falsificationism and verificationism becomes somewhat illusionary.

J. Agassi, a former student of Popper's, explicitly says that the master has changed his views and criticizes him, defending the original version.[17] When commenting on Popper's postulates that the theory resists for some time all attempts at falsification, Agassi writes: "... either Popper assigns no value to positive evidence, *qua* positive evidence [and then the introduced criteria of comparing theories are only an ornament], or he is in the same boat as the inductive philosophers..."[18] Replying to this remark, Popper states: "[Agassi]... disagrees with me about the third requirement which, as he explained to me, he cannot accept because he can regard it only as a residue of verificationist modes of thought.... I admit that there may be a whiff of verification here; but this seems to me a case where we have to put up with it, if we do not want a whiff of some form of instrumentalism that takes theories to be mere instruments of exploration."[19] Criticizing modifications which Popper himself had introduced to his theory, Agassi states directly: "The main problems concerning science, the problems of demarcation and of induction, are solved by the very idea of Popper's program, and are indeed solved in his classical *Logic of Scientific Discovery* prior to any discussion of degrees of testability, of the empirical basis of science, or of corroboration." [20]

Taking into account the fact that the modifications referred to above are not as meticulously developed as the ideas presented in the *Logic...*, as well as the fact that regardless of their value they cannot be consistently introduced into the position represented by that book, I will defer the discussion of these modifications to the next chapter where I will comment on Lakatos' position as he has elaborated them in detail.

The question of whether Popper, in fact, represents naive falsification (in Lakatos' terms) while Lakatos, for the sake of tradition, tries to interpret his position in a way he judges correct, or whether Popper has indeed changed his opinions on particular questions is, in the context of our discussion, a secondary issue.

Popper undoubtedly belongs among those investigators who are open to self-criticism and the revision of their own views – this is evident if only by the footnotes in his works. We know, however, that even the most perfect theories sometimes sink under the weight of modifications that have been introduced into them: I would think that we are here witnessing that phenomenon. And, therefore, I have preferred to present here the falsificationist position in its original version. [21] This is also useful, we note, as this is the form under which it has most often been criticized and analyzed in the literature.

V

Popper's conventionalism differs in important respects from that of Poincaré or Duhem. First of all, if their position can be qualified as conventionalism 'from above', then Popper's is a conventionalism 'from below'; conventions decide as to the acceptance of basic statements.

Secondly, Poincaré assumes that if a theory has been successful for a long period of time then scientists, by means of conventional methods, do not allow for its refutation – they introduce *ad hoc* hypotheses which safeguard it from falsification. According to Duhem, on the other hand an experiment cannot ever serve to refute a theory. They crumble under the weight of the introduced modifications which deprive them of their original simplicity. In assuming such a position, the maintenance or rejection of a theory depends on subjective judgment of its simplicity and not on any objective methodological criteria. The conventions which Popper describes serve not to protect a theory from refutation but, on the contrary, their acceptance is imperative for its elimination. It may then truly be said that this is a difference between 'conservative' and 'revolutionary' conventionalism. "The *conservative conventionalist* (or methodological justificationist, if you wish) makes unfalsifiable by *fiat* some (spatio-temporally) universal theories which are distinguished by their explanatory power, simplicity or beauty. Our *revolutionary conventionalist* (or 'methodological falsificationist') makes unfalsifiable by *fiat* some (spatio-temporally) singular statements..." [22] which may serve as a basis for the falsification of a theory.

Let us notice, however, that in spite of these differences, the 'revolutionary' conventionalist needs conventions which delimit tested theories from their background knowledge. The necessity of this delimitation, that is of a convention concerning the unquestionability of the background knowledge, brings him close to his 'conservative' colleague. Although this convention is to provide the possibility of refuting the theory under test, at the same time it defends the background knowledge from any falsification and, therefore, fulfills a conservative point: *in order to solve one problem, one must accept another as solved on the grounds of some convention.*[23]

Even if Popper's position were modified in the way Lakatos presents it, and the falsifying procedure would be interpreted as a confrontation with the experiment not only of one theory (in conjunction with background knowledge), then we would still be left with the conventions pertaining to the choice of basic statements. It is for this reason, presumably, that Lakatos says: "[We can] try at least to *reduce* the conventional element in falsificationism (we cannot possibly eliminate it)."[24] One way or another, it appears that what makes the system of knowledge 'closed' (fully coherent and justified) are, according to conventionalism, certain conventions. The difference between various versions of conventionalism concerns only the problem of where the conventions enter: on the level of observational (basic) statements or on the level of the choice of theoretical statements or may be on both of them. The real question, how-ever, is whether the system of our knowledge may indeed be coherent and fully justified.

The essence of the problem which faces the conventionalist is the following: science in general, and particularly scientific theories, are to be composed of justified statements. It is impossible, however, to justify everything as we then fall either into a *regressus ad infinitum*, and attempt to solve the problem of what came first, the chicken or the egg, or into a vicious circle, and like the proverbial Baron von Münchhausen, we try to pull ourselves out of the mud by our own ears. The solution to this problem, and there can be no doubt that science does manage to steer around it, must be based, in his opinion, on the assumption that certain statements are conventional in character, that is that we decide as to their veracity on the basis of convention.

Let us reflect, on whether the conventionalist solution is the only anti-dote for the radical empiricist position for which this problem does not exist, because it assumes that observational statements do not require any further justification. The content of sense-data establishes their sense and the conditions of their veracity and, therefore, the refutation of a state-ment in a specific experimental situation would be equivalent with violation of its sense (of semantic rules).

VI

This last remark requires some comment. It could be argued that the radical empiricist is not at all free from the problem which faces the conventionalist. If the refutation of a basic statement in a particular experimental situation is equivalent to the violation of its meaning, that is, of some semantic directives of the language, then it is always possible to change these directives accordingly and to refute the statement. And *vice versa*: if on the ground of semantic directives it is impossible to accept an observational statement, the appropriate modification of these rules would allow us to accept it. Thus, it could be said that a radical empiricist also has to accept certain statements, namely terminological rules, on the ground of conventions.

When speaking of conventionalism, we do not have in mind, however, this trivial view which states that the meaning of natural-language terms is not precise, and that in science we have either to determine it by terminological conventions or introduce into the language of scientific theory some terms which are not current in natural language, i.e., that we must accept by convention some semantic rules and construct artificial languages. He who claims that temperature scales are chosen conventionally and that, therefoie, the acceptance of an experimental report concerning the temperature of a body depends upon this convention, is not necessarily *eo ipso* a conventionalist. Somebody who would believe that, in fact, there exists a uniquely correct temperature scale corresponding to nature itself would have to believe that there exists a pre-established harmony between the universe of things and the universe of signs, i.e. that semantic relationships are strictly determined once and for all, without any human assistance, and that the isomorphism of operations on things and signs is granted *a priori*. As language, however, is a human product and not a mirror for nature, therefore, in order for it adequately to fulfil its communicative and cognitive functions we have to introduce conventional semantical rules into it.

It is important to remember that such a procedure can never be fully performed. Language as a whole must always contain certain terms which are either undefined or, at least, not unequivocally defined. In other words, natural languages constitute a necessary condition for the existence of artificial ones, which are constructed on their basis. Artificial languages, with precisely determined semantic rules, are islands floating on the ocean of natural languages. The link between natural and artificial languages can never be completely severed. The unity of science, which in fact is a *multitude of languages*, can never be achieved; it would be possible only if we could connect all these languages into a coherent linguistic system corresponding to experimental data.

The acceptance of an observational sentence depends then on the semantic rules of language, on the conceptual apparatus, and these rules are established by convention. However, the acceptance of this apparently undeniable view would be equal to accepting the conventionalist position only under the condition that the term 'conceptual apparatus' would be understood as Ajdukiewicz did in the thirties, namely as "a class of all meanings assigned to utterances belonging to a particular *closed* and *coherent* language." [25] Only in this case, if the language of science were at least approximately something like a *closed* and *coherent* language, would the change of the conceptual apparatus imply a change of the whole world-perspective. If, however, the language which we utilize is not *closed* and *coherent*, then a change of the conceptual apparatus leads only to the fact that the same judgment is expressed in different terms, and the new formulation is *translatable* into the old language. The world-perspective does not have to undergo any change. It was precisely the realization that the concept of 'closed' and 'coherent' languages, which are mutually untranslatable, is nothing but 'words on paper', which prompted Ajdukiewicz to abandon the conventionalist position.[26]

In order to illustrate the difference between the thesis that the meaning of certain terms is established by convention, and the point of view of conventionalism, we shall resort to an example.

Let us suppose that we are to solve the problem of whether two spatial intervals are of the same length. We may accept as a unit of measure any calibrated measuring rod. The result of our measurement will, of course, not depend on what scale we may choose to use, be it meter, foot, league, or any other unit. The choice of this scale will have no effect on the solution of our empirical problem. In order to perform the measurement, however, we must transport the conventionally chosen measuring rod from one object to the other. Therefore, the condition of the reliability of the measurement must be the presumption that during the transport the rod will not be deformed in any manner, that its length remain constant, that it remain congruent with itself. According to conventionalism, *this* type of convention is necessary for the solution of an empirical problem. This is, of course, a totally different assumption from one concerning the choice of the measuring scale unit. For whatever scale we may wish to choose, we confront the same situation. A. Grünbaum[27] is undoubtedly right in stating that if we fail to perceive the difference between such conventional decisions we would be compelled to agree, for example, that Einstein, proposing the definition of simultaneity, determined only the semantic sense of the term and not that he has proved that no matter in which way we would define the simultaneity of two events, it remains relative, and that in order to assert their simultaneity we need not only a definition of this term but some *empirical* assumptions concerning measuring instruments. In the case of simultaneity, it is the assumption about the isochrony of clocks.

The dispute with conventionalism does not consist in denying the obvious fact that the meaning of certain terms is established in science by convention. It concerns the problem of whether, in order to solve certain empirical problems, we really do need such conventional assumptions as those pertaining to the behavior of measuring rods or clocks.

VII

Conventionalism, then, is a doctrine according to which some empirical problems can be solved only if we accept the experimental data together with some empirical statements asserted as true by convention. Since the term 'conventional assumption' may be understood at least in two different ways, and each understanding changes the sense of the doctrine, the polemic with it depends largely upon the interpretation of this term.

First of all, a conventional assumption may be interpreted as a statement the truth-value of which can be asserted only by convention. It is the understanding which we find in Poincaré's *Science and Hypothesis*, when he refers to the conventional character of geometry or of the principle of inertia.[28]

Secondly, a conventional assumption may be understood as a statement which functions as a convention in a specific context of inquiry or argument.

I believe that conventionalism is incompatible with empiricism only if the first interpretation is accepted. It seems that even in respect to the second interpretation the term 'conventionalism' is rather misleading.

The claim that a statement functions as a convention means that in the context of inquiry or argument its truth value is asserted by convention. But it does not mean that it can be asserted only by convention in *any* context of inquiry and that we arrived at it by agreement. It is one thing to state that a statement is a convention, i.e., that we have to accept it in the same way as we accept definitions determining the meaning of expressions, and quite another thing to claim that we have decided to treat it as a convention in the given context of inquiry, i.e. to treat it *in this context* as true by definition. The statement 'all metals are conductors of electricity' may be treated either as a synthetic universal statement, or as a basis for stating whether the given substance is, or is not, a metal. In the first case, we are ready to abandon the statement on the grounds of some empirical facts; in the second case, such empirical facts cannot exist at all: every substance which is not a conductor of electricity will be considered *ex definitione* as a non-metal. In these cases the logical status of the sentence is different. Only in the latter may we say that it functions as a convention or as a rule of inference which allows us to infer the statement '*x* is not a metal' from the premise '*x* is not a conductor of electricity'.

In order to clarify the distinction we are speaking about, let us refer to Quine's well known metaphor:

> The lore of our fathers is a fabric of sentences. In our hands it develops and changes, through more or less arbitrary and deliberate revisions and additions of our own, more or less directly occasioned by the continuing stimulation of our sense organs. It is a pale grey lore, black with fact and white with convention. But I have found no substantial reasons for concluding that there are any quite black threads in it, or any white ones.[29]

He who believes that all the statements of science are empirical, perceives only the black threads in the fabric; he who believes that some of the statements are conventions, supposes that there are some very white threads in the fabric; he who believes that the same statement may function as empirical as well as conventional, refuses to accept the empirical-conventional dichotomy as absolute, and treats it as relative, depending on the context of inquiry. The very fact that this third solution is not excluded proves that the alternative between radical empiricism and conventionalism is not exhaustive. It would be exhaustive only if the language of science were a language in the strict sense of the word. But in this case the difference between conventions and empirical statements would be an absolute one, at least synchronically, since the evolution of this language could change the status of particular statements. If, however, science is not such a unified linguistic system, if it is composed of different languages, then a sentence which functions as an empirical statement in the framework of one theory may function as a convention in the framework of another. It is evident that such a possibility would be excluded within the framework of one linguistic system because it would be equivalent to a vicious circle in definitions. The conventionalist perceives just this danger of a vicious circle, and he cannot help but accept the view that some statements are white threads in the fabric of our knowledge.

He would agree, of course, that the language of science evolves as a whole. He is ready to assert, as did Poincaré, that "When a law has received a sufficient confirmation from experiment... we may elevate it into a *principle* by adopting conventions such that the proposition may be true.... The principle, henceforth crystallised, so to speak, is no longer subject to the test of experiment."[30] Therefore, at a certain stage of evolution of science (and of its language), the sentence achieves the status of a convention. Or, he would state, as Ajdukiewicz did, when he still held the conventionalist position, that prior to Newton the language of science contained no such semantic rule which would have prohibited questioning the sentence 'when a force exerted on a body is not neutralized by an equal force, the body changes its velocity', while after Newton the language has changed in such a way that 'it contains an axiomatic semantic rule pertaining to this sentence.'[31]

This signifies, of course, that at this new stage Newton's third law became a conventional semantic rule which determined the meaning of the term 'force'.

It is one thing, however, to claim that the status of some sentences of the language of science changes *historically*, and quite another to acknowledge that a given statement may fulfill the function either of an empirical sentence or of a convention within the same period of evolution of science. The first claim expresses the opinion that science is always a system of knowledge whose coherence is granted by conventions, that the distinction between 'black and white threads' is synchronically absolute even though the language of science evolves. At least some empirical problems may be solved only by convention. The second claim states that the solutions of empirical problems depend not only upon the experimental data, but also upon the statements of other theories and upon the whole of our knowledge, which we utilize in the solving of our problem. The fact that we treat some of the claims of the background knowledge as conventions does not mean that they were accepted by convention within other theories.

The conventionalist does not maintain that the language of science does not change and that the same statements always function in it as conventions. He does claim, however, that at a given stage of the evolution of knowledge, the division between conventions and empirical statements is absolute. If we do not perceive this fact, we run the risk of an inadequate criticism of conventionalism. The conventionalist claims: we perform measurements in science, but in order to do so, we must refer to certain laws; and they, in turn, are formulated on the basis of measurements. We find ourselves in the grip of a vicious circle, and in order to free ourselves from it, we must assume that certain statements which we utilize are true by convention. In order to formulate the law of thermal expansion, it is necessary to utilize the thermometer, but in order to construct a thermometer it is necessary to refer to the law of thermal expansion. This conventionalist claim does not imply, however, as some critics argued, that the world must have been endowed with the thermometer from the time of creation. This argument would be conclusive only against such a version of conventionalism which would claim that the same sentences always function as conventions, that the structure of the language of science does not change. I have never read any text in which such a thesis would have been defended. The works of Poincaré and Ajdukiewicz show that the conventionalist claims that certain statements, empirical at their origin, may become conventions and serve as basis for the interpretation of experimental data. It is just this thesis which is the subject of controversy.

Only if the language of science were unified into one system at some point of its development would an absolute distinction between empirical statements and conventions be justified. However, in the strict meaning of the term, only fragments of our knowledge, i.e. particular formalized theories can be considered as languages. They never function autonomously: in order to interpret the results of experiments they must refer to other theories and to everyday language. Statements borrowed from other fragments of our knowledge are not and cannot be justified in the frameworks of these theories. Therefore, it is only in respect to these theories that the acceptance of them can be conventional. These are 'the grey threads' with which we weave our knowledge into a whole. This whole is always provisional and can be torn apart under the pressure of new experience. We weave it in such a way as to avoid contradictions within the system which, in fact, is never completely coherent.

This may sound like a paradox, but it seems that if the program of the unity of the language of science were ever realized, we would be forced to acknowledge that certain scientific statements are conventions in the absolute sense. But it is obvious that such a language would have to include not only a theory of the universe, but also a theory of experiencing and cognition, not only the methodological rules but rules of application of these rules as well, and we do not have to worry about this possibility.

The cogent conviction that no hypothesis is ever tested experimentally in isolation but always in connection with some background assumptions, leads the conventionalist to believe that some of these statements must be asserted as true conventionally. He believes that this is the only way for the achievement of a fully coherent and justified system of knowledge. It seems, however, that this belief is an illusion.

For it is an illusion to assume that we are able to specify all the assumptions we accept when we are testing a hypothesis. Some of them become evident only after a competing theory appears. Our knowledge could have been a coherent and fully justified system only if we were able to point out *all* the assumptions, which we have accepted, *before* a new 'crazy idea' appeared and questioned the beliefs that all scientists accepted without even supposing that they are not as obvious as they believed them to be. It seems that if scientists did not accept assumptions which, no matter whether wrong or right, are questionable, they would be unable to formulate any theory. The acceptance of such assumptions seems to be a necessary condition for the existence of science. Without them there would be no islands of fully justified systems of knowledge on the ocean of common-sense knowledge. So, I would say that it is common-sense knowledge rather than conventions which constitutes the irremovable crutches on which science rests. I do not see any reason why all these assumptions on which scientific knowledge rests, and which we test together with our theories, often *without even knowing what we assume*, are to be treated as conventions.

It seems important to stress that the thesis that the conceptual apparatus co-determines the empirical facts should not be understood as equivalent to the thesis that these facts are co-determined by the language of science, whether in the strict or in the common sense of this term. The above theses are not equivalent because, first of all, the empirical facts are co-determined not only by the conceptual apparatus of the theory which explains these facts, but also by the conceptual apparatus of the background knowledge. And it is impossible to believe that all this conceptual apparatus can constitute a language in the strict sense of the term. This is why I believe Popper is right when he states: 'I do not think that the study of the growth of knowledge can be replaced by the study of linguistic uses, or of language systems.'[32]

Moreover, if we accept that the background knowledge co-determines the empirical facts, we neither have the possibility to restrict its scope (for example, only to accepted scientific theories), nor to specify all the assumptions which we have accepted before they are questioned by a new theory. Thus, the term 'language of science', even if understood in the strict sense of the term, is not appropriate to express what we have in mind when we state that the conceptual apparatus we use co-determines the meaning of scientific claims and has an impact upon their acceptance or refutation. What determines the meaning of empirical facts is a conceptual apparatus of various scientific theories which cannot always be reduced to each other, as well as a conceptual apparatus of common sense knowledge.

Only if we delimit science from commonsense beliefs by means of some methodological criteria which we accept consciously or unconsciously, only if we separate it from the breeding-ground on which it grows, are we obliged to accept the view that some of its claims are conventions. It is then that the mistaken belief according to which science is treated as a well-delimited system of fully justified statements finds its expression in the conventionalist philosophy. And then, philosophy of science, which aspires to explain the growth of knowledge, must appeal to some conventional rules. It needs them always, whenever it has to cope with changes which have their source outside of a system delimited in this manner.

NOTES

[1] "It is often said that experiments should be made without preconceived ideas. That is impossible. Not only would it make every experiment fruitless, but even if we wished to do so, it could not be done. Every man has his own conception of the world, and this he cannot so easily lay aside. We must, for example, use language. And our language is necessarily steeped in preconceived ideas. Only they are unconscious, preconceived ideas which are a thousand times the most dangerous of all." (H. Poincaré, *Science and Hypothesis*, New York 1952, p. 143.)

[2] P. Duhem, *Aim and Structure of Physical Theory*, New York 1962.

[3] K. R. Popper, *Conjectures and Refutations*, London 1969.

[4] Existential universal hypotheses are unfalsifiable. That is why basic statements must have the character of temporarily and spatially restricted existential hypotheses.

[5] K. R. Popper, *The Logic of Scientific Discovery*, pp. 101–102.

[6] *Ibid.*, p. 86.

[7] Our example is, of course, oversimplified; the hypothesis is not a universal statement.

[8] R. Carnap, *Logical Foundations of Probability*, Chicago 1951; H. Reichenbach,

Experience and Prediction, Chicago 1938; *The Rise of Scientific Philosophy*, Berkeley 1951. Cf. also S. Amsterdamski, 'O obiektywnych interpretacjach pojęcia prawdopodobieństwa' (On the Objective Interpretations of the Concept of Probability) in *Prawo, konieczność, prawdopodobieństwo*, Warsaw 1964, pp. 1–125.

[9] The possibility of the probabilistic estimation of scientific hypotheses is a matter for discussion. First of all, it may be doubted whether the Carnapian theory provides the possibility of a quantitative estimate of the degree of confirmation when the evidence, in respect to which we estimate the probability of a hypotheses, is not statistical. (S. Amsterdamski, *op. cit.* Part I). Secondly, even if this estimation is possible, it is irrelevant in respect to the truth of the estimated hypothesis. A false hypothesis which follows deductively from false premises has the probability $p=1$. (Cf. S. Amsterdamski, *ibid.* and 'Prawdziwość i prawiopodobieństwo' (Truth and Probability), *Zeszyty Naukowe Uniwersytetu Lodzkiego*, 1967, pp. 1–15). Finally, it is doubtful whether the Popperian conception gives the possibility of a quantitative estimation of a hypothesis, since we can never point to all its potential falsifiers. Accepting the Popperian theory of probabilistic estimation, we would be compelled to state that the general theory of relativity hardly satisfied his methodological requirements, as we can point out few potential falsifiers, while the general field theory does not satisfy them at all. I am not convinced whether his and the Carnapian theories of the probabilistic estimation of scientific hypotheses are as opposed as Popper believes. It seems rather that both of them attempt to apply the calculus of probability for different aims, and are based on different interpretations of the calculus. The practical applicability of both conceptions in scientific research seems very limited.

[10] The detailed analysis of these conventions may be found in I. Lakatos, 'Falsification and the Methodology of Scientific Research Programmes', in *Criticism and the Growth of Knowledge* (ed. by I. Lakatos and A. Musgrave), Cambridge 1970, pp. 91–195.

[11] I. Lakatos, *ibid.*

[12] T. S. Kuhn, 'Logic of Discovery or Psychology of Research', in *Criticism and the Growth of Knowledge* pp. 1–27.

[13] N. Bohr: 'Discussions with A. Einstein on Epistemological Problems in Atomic Physics', in *Albert Einstein – Philosopher Scientist* (ed. by P. Schilpp), Evanston 1959, pp. 199–243.

[14] K. R. Popper, *Conjectures and Refutations*, pp. 215–248.

[15] *Ibid.*, pp. 242–248.

[16] *Ibid.*, pp. 231–234.

[17] J. Agassi, 'Science in Flux,' *Boston Studies in the Philosophy of Science* (ed. by R. S. Cohen and M. W. Wartofsky), Vol. III, Dordrecht 1967, pp. 293–324.

[18] *Ibid.*, p. 311.

[19] K. R. Popper, *Conjectures and Refutations*, p. 248, footnote 31.

[20] J. Agassi, *ibid.*, p. 314.

[21] In a footnote to the chapter 'Unity of Method' in *Poverty of Historicism*, Popper wrote (p. 131): "See my *Logic of Scientific Discovery* on which the present section is based, especially the doctrine of test by way of deduction ('deductivism') and of the redundancy of any further 'induction', since theories always retain their hypothetical character ('hypotheticism'), and the doctrine that scientific tests are genuine attempts to falsify theories ('eliminationism')...".

[22] I. Lakatos, 'Falsification and the Methodology of Scientific Research Programmes' in *Criticism and the Growth of Knowledge*, p. 106.

[23] The conventions postulating the unquestionability of the background knowledge

would be needless only if both of the competing theories were linked with the same background assumptions. Lakatos seems to accept this premise implicitly, though it is by no means unquestionable.

24 I. Lakatos, *ibid.*, p. 115.

25 K. Ajdukiewicz, 'Das Weltbild und die Begriffsapparatur', *Erkenntnis* **4** (1934) 262.

26 K. Ajdukiewicz, 'W sprawie artykułu profesora A. Schaffa o moich poglądach filozoficznych' (Reply to Professor A. Schaff's Paper on my Philosophical Opinions), in K. Ajdukiewicz, *Język i Poznanie* (Language and Cognition), Vol. I, Warsaw 1965, p. 177.

27 A. Grünbaum, 'Geometry, Chronometry and Empiricism', in: *Minnesota Studies in the Philosophy of Science*, Vol. III, 1062, pp. 405–527 (cf. especially pp. 419–424).

28 H. Poincaré, *Science and Hypothesis*, New York, 1952: "...no experiment will ever be in contradiction with Euclid's postulate; but on the other hand, no experiment will ever be in contradiction with Lobatschewsky's postulate" (p. 75) and "To sum up, whatever way we look at it, it is impossible to discover in geometric empiricism a rational meaning." (p. 79) And further: "...this law [of inertia] ...can be extended fearlessly to the most general cases; for we know that in these general cases it can neither be confirmed nor contradicted by experiment." (p. 97)

29 W. V. Quine, 'Carnap and Logical Truth', in *Logic and Language – Studies Deducated to Professor Rudolf Carnap*, Dordrecht 1962, p. 63.

30 H. Poincaré, *The Value of Science*, New York 1958.

31 K. Ajdukiewicz, 'Das Weltbild und die Begriffsapparatur', p. 263.

32 K. R. Popper, *The Logic of Scientific Discovery*, p. 16.

CHAPTER VI

REFORMATION AND COUNTER-REFORMATION: PARADIGMS AND RESEARCH PROGRAMS

I

In the last two chapters, we had the opportunity of inspecting the problems faced by a philosophy of science which attempts to explain the process of the development of scientific knowledge in logical or methodological terms.

This explanation cannot be provided by radical empiricism which is motivated by the mistaken conception of the unilateral relationship between facts and theories. The epistemological impossibility of delimiting purely empirical facts and the resulting methodological impossibility of differentiating between observational and interpretive statements compels us to reject this over-simplified scheme of the development of science, according to which the evaluation of scientific theories (that is their acceptance or rejection) is determined exclusively by an autonomous empirical basis.

Popper's falsificationism, on the other hand, gives up this conception and treats all scientific statements as theoretical and, therefore, questionable. It perceives that the acceptance and rejection of observational sentences depends not only on the pure results of experiment but also on their interpretation in terms of contemporary background knowledge; but it is in trouble when on this basis it seeks to reconstruct the process of the development of knowledge, or more precisely, to establish methodological rules, the application of which determines this historical process.

The history of science well demonstrates that the replacement of old theories by new ones does not proceed according to the falsificationist scheme. First of all, the acceptance of an observation statement which contradicts a hitherto accepted theory does not always lead to the refutation of that theory. Secondly, the falsification of the old theory is not a necessary condition for the acceptance of a new one. Thus, while the falsificationist model of the evolution of science is 'richer' than the radical empiricist model as it takes into account the effect of already acquired knowledge on the further process of evolution, nevertheless it is still not rich enough, for it fails to take into account certain important factors which co-determine the development of scientific knowledge, particularly in the periods of its fundamental transformations, i.e. of revolutions.

As I have attempted to demonstrate, falsificationism cannot take into account the role of these factors because it accepts the following assumptions: (1) the restriction of philosophical reflection on science to the problems of logic (methodology), as a result of which historically variable factors are either totally excluded from the scope of investigation or, when it becomes impossible to ignore them, their analysis is replaced by the introduction of methodological conventions; (2) the delimitation of the field of science by means of an accepted criterion of demarcation; in consequence, the historically variable methodological rules which are a product of contemporary intellectual culture or, more precisely, of the function of science within this culture, are replaced by supra-historical criteria of rationality which constitute the basis of logical reconstruction; (3) in strict connection with the preceding assumptions, the understanding of the development of science as an autonomous process, advancing according to immanent regularities which find their expression in supra-historical methodological rules; this must lead to the treatment of all other factors of scientific development as inessential and to the splitting of the history of science into 'internal' and 'external' history.

The difficulties which we have mentioned above have been noted by those investigators who in their studies on science have referred to such categories as 'the style of scientific thinking' or 'paradigms'. These concepts, it appears, have not yet been elaborated with satisfactory precision. Different factors, such as philosophy, value system, tacit knowledge[1], the state of knowledge in the given field and in related disciplines, particularly the state of mathematics, the organization of scientific life, social needs, the situation of science as a social institution – are treated as components of the cultural heritage which affect the growth of knowledge. The role of each of these factors and their mutual relations are differently appraised by individual investigators. What is common, however, to all these conceptions, is the awareness that the present state of the philosophy of science cannot adequately account for the process of the evolution of science, and hence a search for a way in which to improve this situation.

In this chapter, we shall discuss the well-known recent proposals, presented by Kuhn in *The Structure of Scientific Revolutions*, for the reform of the traditional approach to the history of science and its methodology.

We shall then deal with the modifications of the falsificationist model of the development of knowledge proposed by Lakatos – undoubtedly under the influence of Kuhn's critique of Popper. Confrontation with these two positions will allow us to set forth certain theoretical problems which in turn will be analyzed in the last two chapters.

II

Kuhn assumes that neither radical empiricism nor falsificationism can provide a satisfactory description of the mechanism of the development of science. Therefore, he attempts to change the traditional method of approaching history as well as philosophy of science.

> Today research in parts of philosophy, psychology, linguistics and even art history all converge to suggest that the traditional paradigm is somehow askew. That failure to fit is also made increasingly apparent by the historical study of science to which most of our attention is necessarily directed here.
>
> None of these crisis-promoting subjects has yet produced a viable alternative to the traditional epistemological paradigm, but they do begin to suggest what some of the paradigm's characteristics will be.[2]

The basic concept of Kuhn's book is undoubtedly the concept of the paradigm, which is to be the means for overcoming the aforementioned crisis. According to Kuhn, the paradigm determines the practice of normal science, specifying which questions are to be investigated and providing criteria for the acceptance of the results of these investigations. It constitutes, therefore, a kind of 'point of view' from which one undertakes and conducts investigations or, as Kuhn notes, a manner of seeing the world, and contains the 'rules of the game' called science.

The development of science proceeds through three stages: (a) the development and application of a given paradigm, that is the resolution of problems which it poses in a manner which it sanctions; (b) a period of crisis or revolution, when the hitherto accepted paradigm encounters problems, which it itself has brought to light but which it cannot solve, a period during which various attempts at resolving the problem vie with each other but during which a new, widely accepted paradigm is still lacking; and (c) the overcoming of the crisis and the rise of a new, universally accepted method of research – in other words of a new paradigm. And then the cycle repeats itself.

The period of a scientific revolution when there is no reigning paradigm and, therefore, no universally accepted criteria for the evaluation of hypotheses is a point of discontinuity in the process of growth of knowledge. Old criteria are no longer binding, scientists have lost their trust in them, and new ones have not yet been developed. In order to understand what a scientific revolution is, it is necessary to investigate how the new paradigm replaces the old. Since paradigms contain specific rules for the acceptance and rejection of scientific statements, and since there is no 'neutral', extra-paradigmatic arbitrator in the competition between two paradigms, the logic of scientific discovery cannot reconstruct this process of transition. "[The] methodological drectives, by themselves [are not sufficient] to dictate a unique substantial conclusion to many sorts of scientific questions"[3] with which scientists must deal during the process of a scientific revolution.

Kuhn's description of scientific revolutions brings to mind Plato's allegory of the cave. Just as the slaves, bound to the rock, had to take appearance for reality until one of them could break his bonds and exit from the cave to announce to his companions that they see only shadows and not the real things, so those who are bound by the accepted paradigm are unable to doff it until someone from the outside, a young scholar or someone just beginning his work in a given discipline and not shackled by tradition, may look at the facts from another point of view and convince them that it provides a better cognitive perspective. There is, however, one important difference in this analogy. According to Plato, the one who discarded the chains and announced to his fellow prisoners that they saw only shadows of real objects was to have revealed an absolute truth. According to Kuhn, one who proposes a new paradigm presents a proposal which is as relative as the old one. He proposes that we simply move to a different cave. In certain respects, the new paradigm is no different from the old one; like the old one it will be self-confirming: it will be impossible to reject it from the inside by means of logical criticism; just as its predecessor it will impose a particular way of seeing the world. It is precisely as a result of the very nature of paradigms, which are like closed and mutually untranslatable languages, that the successive points of view are

at least partially incommensurable.[4] This conception of the mutual relationship between successive paradigms, which provide different views of
the world, closely parallels Ajdukiewicz's theory of *closed* and *coherent*
languages which we had occasion to mention in our previous chapter.
There is, however, one important difference between these two theories –
Ajdukiewicz's theory was presented with a much larger degree of precision than that of Kuhn. This lack of precision has, as always, two consequences; it aids in the defense of the theory and makes it that much more
difficult to criticize.

Kuhn's conception of the transition from one paradigm to another,
which we have briefly presented here, raises two essential questions. First
of all, it contests the continuity and cumulative character of the development of knowledge; and second, it questions the rational character of
this process. Thus, the following question arises: if we really are prepared
to reject the conception of supra-historical criteria of rationality, are we
then bound to accept these conclusions? We shall come back to this question in the next chapter.

III

As we have noted, the key concept in Kuhn's theory is the concept of the
paradigm. Appraising the value of the proposals presented in his work, it is
impossible to deny the fact that this term is rather ambiguous. In an essay
which is otherwise quite sympathetic to Kuhn's theory, M. Masterman
has enumerated twenty-one different ways in which he uses this term.[5]

As is obvious from my presentation, in many respects I appreciate the
Kuhnian approach. It immediately raises questions pertaining to factors
which condition the evolution of science but which fall outside the realm
of the philosophy of science understood as limited to methodological
problems. Instead Kuhn raises the problems of the tradition of practicing
science, of the transmission of the tradition from masters to their students,
of the role of textbooks and classical monographs as factors which delimit
the scope and methods of scientific research. Simultaneously, he rejects
the concept of supra-historical rules or criteria of 'rationality' as they are
defined by the logic of scientific discovery and seeks them within historically determined methods of practicing science in a given period, that is
within the frames of the accepted paradigm. The change of a paradigm
is, at the same time, a change of these criteria. These ideas are undoubted-

ly valuable and are responsible for the widespread interest in *The Structure of Scientific Revolutions*. Surely Kuhn's book has been one of the most often quoted works within the philosophical and historical literature devoted to the problem of the development of knowledge within the last decade.

Despite all these merits, it is difficult to deny that Kuhn's concept of paradigm plays in this book the role of the proverbial bottomless well, into which one may cast various factors which, although they undoubtedly condition the evolution of knowledge, cannot be theoretically encompassed under one category. The ideographism from which Kuhn rightly wants to free the history of science as well as the vulgar empiricism of which he, not without reason, accuses many contemporary theories of the evolution of knowledge, come back in his own work, through the back door so to speak. The historically variable factors which condition the development of knowledge, relegated by the philosophy of science to 'external history' are, admittedly, rehabilitated as essential for the understanding of this process. Nevertheless, they are contaminated by the concept of the paradigm rather than synthesized by it. The root of this problem, as I see it, is to be found in the particular concept of science with which Kuhn approaches the study of its evolution, i.e. the concept of *normal* science. Construing the notion of paradigm in such a manner as to correspond to his notion of normal science, he is forced to include in it various heterogeneous factors which do not constitute a coherent whole. Thus the notion of paradigm will embrace such elements as laboratory equipment, textbooks, monographs describing patterns for the solution of scientific problems, as well as the sociological description of the scientific community and ontological assumptions pertaining to the structure of the universe, etc. Although it is the paradigm that is supposed to determine the way of conducting normal scientific research, in reality Kuhn's understanding of this concept results from his understanding of normal scientific activity. In the *Postscript* to the second edition of his monograph, he admits that his understanding of the paradigm and of normal science leads to a vicious circle: the paradigm is the manner of conducting normal science – normal science is scientific activity based on a common paradigm. Kuhn sees the solution to this dilemma in the possibility of a sociological characterization of scientific communities, and in the construction of the concept of paradigm on the basis of an analysis of the beliefs held in common by the members of these communities. This, I think, confirms our opinion that Kuhn's notion of normal science is prior to his concept of the paradigm.[6] Let us investigate this notion further.

As opposed to Popper, for whom scientific activity is a permanent revolution, Kuhn, and not without good reasons, states that scientists spend most of their time and effort not in overthrowing accepted theories but in solving problems which these theories have raised. In effect, they do not subject a theory to constant attempts at falsification, but rather try to extend it and render it more precise. What, according to Popper, is an everyday situation in science, according to Kuhn occurs only exceptionally, when the lack of a reigning paradigm forces scientists to severely screen candidates for this role and to subject them to rigorous attempts at falsification.

Thus, it is not the tendency towards falsification of theories, but rather towards the solution of various problems – or as Kuhn calls them puzzles – which they raised which constitutes the essence of scientific activity. The specific characteristic which isolates science from other fields of human activity is in fact just this tradition of solving such kinds of problems, and not the attempts at overthrowing theories.[7]

Kuhn's concept of normal science has been severely criticized by various commentators on his book.[8] They have raised the objection that situations where there has been only one accepted paradigm are rather the exception and not the rule in science,[9] as well as the objection that Kuhn's concept of normal science sanctions dogmatism in science. Commenting on this point, Popper wrote:

I admit that this kind of attitude exists; and it exists not only among engineers, but among people trained as scientists. I can only say that I see a very great danger in it and in the possibility of its becoming normal (just as I see a great danger in the increase of specialization, which also is an undeniable historical fact): a danger to science and, indeed, to our civilization. And this shows why I regard Kuhn's emphasis on the existence of this kind of science as so important.[10]

Regardless of how just these remarks might be (the first I consider undoubtedly correct, the second, however, is quite debatable as it is not absolutely apparent that in speaking of normal science, Kuhn is describing a particular state which goes beyond the description of an existing phenomenon and considers it as a positive norm for future research, as some of his critics have imputed), it appears to me that the crux of the matter lies elsewhere.

As I see it, when we ask ourselves the question – what is really the object whose development Kuhn is describing? – we come face to face with a troubling problem. When the author speaks of the need for a new philosophical approach to the development of knowledge, he is undoubtedly thinking of the evolution of science as a whole. For the philosophy of science does not deal with the construction of theories of the evolution of physics, chemistry or any other particular discipline of science, but with the evolution of scientific cognition, *tout court*. However, when Kuhn proceeds to the analysis of scientific revolutions, the transitions from one paradigm to another, as a rule he describes the operation of this process within particular disciplines. As a result his revolutions are not revolutions in *science*, but revolutions within *particular disciplines*. By analogy, the concept of normal science does not refer to science in general, but to the manner of development of particular disciplines. At this point, it is difficult to discern what Kuhn really has in mind when he refers to a particular discipline. At certain points he considers them in a wider manner and speaks of revolutions in astronomy, chemistry or physics, at other times he treats them more narrowly and, in effect, discusses them in a more narrow sense; it then appears that even the discovery of X-rays is considered to constitute the breakdown of an old paradigm within a specific narrow specialization.[11] This is immediately mirrored in the concept of the paradigm. At times it includes such factors as the construction of a new experimental device, at other times such general elements as the modifications of the philosophical concepts of the structure of the universe.

It is, of course, possible to reply that there, in fact, is no such thing as science 'in general' beyond specific disciplines and, in a certain sense, this is truly the case. No one is ever engaged in the practice of science in general, and it would be hard to imagine what such an activity could be. If, however, one assumes that the question concerning the criterion of demarcation is not meaningless, that there are some criteria which delimit this set of individual disciplines, if one assumes that this set can be characterised in some theoretical manner, that there exist reasons for denoting all these disciplines by a common term – and Kuhn never subjects this to any doubt, to the contrary he is looking for just such characteristics – then, by the same token, it cannot be denied that there is something which connects and delimits them as a certain whole. In such a case, however, a revolution in science and a revolution in a specific discipline are by no means the same thing.

The critics of the monoparadigmatic concept of the evolution of science have stressed that the monopoly of one paradigm constitutes the exception rather than the rule. Although I am in complete agreement with this view,[12] I would like to raise here a problem of a different nature: the scientist who approaches a particular problem is not led by one paradigm, but by several different 'points of view'. Some of them are more general, others are more specialized. This is not to say that they are all coherent, if only for the reason that such coherence never exists in science as a whole. If paradigms determine the tradition of normal research then this tradition is never as uniform as it would appear from reading Kuhn's book.

Let us suppose that it would be possible to single out certain theses which belong to the paradigmatic assumptions of contemporary physics and are, therefore, accepted by all physicists. There can be no doubt, however, that if we took under consideration not physics as a whole, but let us say quantum mechanics, we would be forced to conclude that specialists within this field hold additional paradigmatic assumptions beside those accepted by all physicists. The narrower the field, the more specialized will be the paradigm. If this is so, then it must be that a discovery, which from the narrower point of view would already not fit within the tradition – the accepted paradigm – and would, therefore, be treated as a revolution, would from another, wider perspective, still lie within the bounds of tradition, would be considered as its continuation, as a successive step in the evolution of normal science. From one point of view, for example, the special theory of relativity marks the breakdown of the classical paradigm of Newtonian physics but from another point of view it can be considered in certain respects as a continuation of its (deterministic, for example) program. It is well known that the partisans of the Copenhagen interpretation of quantum mechanics considered Einstein a dogmatist[13] because he refused to renounce certain paradigmatic assumptions of classical physics such as determinism and the postulate of a completely objective description of natural phenomena.

Therefore, if the concept of a paradigm is defined in a wider manner, then it cannot encompass all those assumptions which guide the work of the majority of scientists in a given period, that is those which are associat-

ed with work in a narrow specialty. If, on the other hand, it is defined in
a narrower, more specialized manner, then it does not determine the tradition of normal science, as this consists of a whole aggregate of paradigms
(paradigmatic assumptions) of various levels of generality, and of various
character. This signifies, then, not only that *not all* investigators are
guided at the same time by the same paradigm, but also that the same
scientist may be guided in his investigations by various points of view, of
various degrees of generalization, which may superimpose themselves on
each other, or fall into conflict with each other. This point is of crucial
importance for the evaluation of Kuhn's concept of revolution as a break
of continuity in the development of knowledge. For it is immediately apparent that a revolution in some specialized field of research does not
necessitate the overthrow of all the paradigmatic assumptions held by
investigators within that specialization, but only of those, or some of
those, which are limited to this narrow field. Therefore, the differentiation
between revolutions in particular disciplines and in science in general
becomes unavoidable and of primary importance. It could be avoided
only in one case, namely if by the term 'science' we assumed a set of autonomous, mutually independent disciplines. In such a case, it is clear,
we could properly speak only of revolutions within narrowly defined,
specialized fields of inquiry. But by that very same token the questions:
'What is the cause of revolutions in science in general?', and 'What is the
cause of revolutions within specific disciplines?' are not completely synonymous. They are not synonymous as they refer to two culturally different phenomena; the first happens in science almost every day, the
other very rarely.

IV

The question arises, why is it that the author of *The Structure of Scientific
Revolutions*, who otherwise cannot be accused of a lack of clearsighted-
ness, does not perceive the ambiguity of the basic category of his whole
concept – the paradigm and the notion of scientific revolution? Is a
paradigm a point of view and a method for solving particular problems
within a specialized field, or a manner of practicing science within a
specific period? And, accordingly, is a scientific revolution a modification
of certain paradigmatic assumptions within a specific narrow field of inquiry, let us say, plant cytology, a modification which can have little or

no effect on the manner of practicing biology as such, and much less science as a whole and, therefore, is purely local in character; or is it a modification of certain paradigmatic assumptions which are accepted in all fields of research in a given period? It seems that behind this impreciseness of terminology, which Kuhn himself admits,[14] something deeper, more fundamental lies hidden.

What, in fact, appears to lie behind this ambiguity is the conviction that the factor which demarcates science from all other forms of human intellectual activity is to be sought in the study of the 'normal' activity of scientists, that is of normal science. And what is characteristic of normal science is the puzzle-solving tradition. And this very tradition may be treated as the criterion of demarcation.[15] Before the rise of the first paradigm within a specific field, we are not dealing with a mature science.

In order to substantiate this position, Kuhn brings into the discussion the example of astrology. Popper has on many occasions specified that by his criterion of demarcation, astrology lies outside the bounds of science. Kuhn agrees that astrology was never a science, but not for the same reasons as Popper. The latter claims that astrological prognoses were deceptive and the assumptions of astrologers unfalsifiable. Kuhn, on the other hand, proposes to compare the activity of the astronomer and the astrologist. When the predictions of the former failed to come true, he could always hope that he could correct this by introducing supplementary hypotheses – new epicycles, deferents, eccentrics – or by following up with more precise observations. This constituted the astronomical tradition of solving puzzles. The astrologer, on the other hand, had no such problems to solve and was never involved in this type of activity. In distinction to the astronomer, his failure did not provide a basis for the posing of new questions and the formulation of new hypotheses. He could at best justify his failure by concluding that we do not know the exact position of the celestial bodies, or the precise moment of birth for the person whose horoscope was being prepared. So despite the fact that astrologers formulated testable hypotheses and realized that for one reason or another they may turn out to be wrong, they were, nevertheless, not involved in the kind of activity which determines the tradition of solving problems. In this situation, even if there were some correlation between the movements of celestial bodies and human destiny, astrology would still not be a science.

Let us accept for the time being Kuhn's explanation and ask another question: to what purpose does a scientist solve puzzles?[16] As Kuhn does not provide us with any direct answer to this question, we must look more closely at what kind of problems he decides to term as the puzzles of normal science: or, in other words: since normal science is to be the basis for characterising science in general, what are its basic features?

First of all, the value of a puzzle does not lie in the importance of its solution, but in the manner in which this solution is obtained.[17]

Second, a puzzle is a problem for which we know that there must be a solution. And it is, in fact, the paradigm which determines which problems can be solved on the basis of a given theory and are, therefore, scientific, and which ones are not worthy of any attention, as they are metaphysical, unscientific, lie within the bounds of another discipline or are not yet mature enough to be the object of investigation.[18]

In the third place, scientists tend to concentrate on problems for which it is known that there is a solution, which is why normal science can advance in a rapid manner. In such cases, failure to solve a puzzle can be caused only by the individual ineptitude of an investigator. The desire to demonstrate one's ability to solve puzzles is in fact one of the essential motivating factors that induces men to undertake scientific research.[19]

Fourth, the solution of a puzzle is to be reached by the application of specific rules which must not be altered, just as in solving a chess problem one is not allowed to move the pieces in an arbitrary manner. The set of these rules is embodied within the paradigm.[20]

Let us remember, moreover, that Kuhn explicitly rejects the concept according to which the development of science is viewed as a progressive evolution towards the truth.[21] (This follows quite apparently from his conception of the paradigm and the transition from one point of view to another in the absence of any supra-paradigmatic instance.) The criterion for accepting a solution lies, according to Kuhn, in the unanimous common opinion of scholars who are directed by a common paradigm. As he himself states it: "What better criterion than the decision of the scientific group could there be?"[22]

It is already apparent from his characterization of problems, which he terms puzzles, and which are the subject for solution by normal science, that the goal of normal science does not lie in the seeking of new facts or theories. On the contrary: "Even the project whose goal is paradigm articulation does not aim at the *unexpected* novelty."[23] What then are the traits of that model of scientific activity, which is to constitute the basis for the characterization of science?

We must digress here in order to provide a certain parallel. In explaining to the U.S. Senate Committee on Atomic Energy in 1945 the difference between fundamental and applied research, Robert Oppenheimer stated as follows:

Dr. Wilson said that programmatic research consists in the measurement of quantities which are known to exist. That may sound very funny, but half of the work of physics and most of the work of fundamental physics, is not to measure the quantities that are known to exist, but to find out what quantities do exist. That is, what kind of language, what kind of concepts correspond to the realities of the world. What Dr. Wilson was saying – and I think quite rightly – was that if you really know what it is you have to do if you can give it a name and say 'We don't know whether it is 10 or 20, but we would like to measure it', then you are doing something that increases knowledge very much but which doesn't qualitatively increase knowledge. When you try to find out whether it is possible to talk about such things as simultaneity and position ... then you are going to make a contribution.[24]

It is difficult to avoid the impression that this characterisation of applied research is very close to Kuhn's conception of normal science. In both cases the success of an investigator depends totally on the demonstration that the accepted theory may be effectively utilized and applied to the solution of the puzzle in question. In both cases what is at stake is not the search for a new way of ordering the universe of human experience, but the strengthening of the already existing order; not the search for truth (whatever this term may signify), but the utilization of truths already achieved.

We have already spoken in Chapter Two of how we view the function of science within human culture. We may now add that both the Popperian vision of science as permanent revolution, and also Kuhn's vision of normal science presented as a model of scientific activity, constitute an absolutization of one side of this function, which has as its role to unify two orders – knowledge of 'how' and of 'why', of what is possible and of what is theoretically impossible. In Popper's view the goal of science lies *exclusively* in the search for truth. Its eyes are solidly directed towards the heavens, but the sky is to be free of any metaphysical clouds. It is to be simply a 'good' philosophy and Popper's methodology is to serve to this aim. Kuhn's normal science, on the other hand, is to consist exclusively in the solving of problems, the solution of which is guaranteed by the accepted paradigm. Its eyes are solidly directed towards earth, but this is an earth free from any metaphysical troubles, as these have already been solved (or buried) by the accepted paradigm. It will indicate to scholars which problems are scientific and decide which are to be "rejected as metaphysical, as the concern of another discipline, or sometimes as just too problematic to be worth the time."[25] It is worth stressing the fact that the demarcation provided by the paradigm is already determined by the philosophical assumptions which it contains. Therefore, if Popper's criterion of demarcation is absolute, Kuhn's is relative. But both have as their purpose to separate scientific investigation from metaphysical problems. In this sense it is possible to state that despite all of Popper's and Kuhn's anti-positivist declarations the argument remains, so to speak, within the family.

It seems, then, that the difficulties of both of these conceptions have a common source, while the conflict between them results from different value-systems which they implicitly accept. The source of the difficulties lies in the underlying solutions to the problem of the relationship between science and metaphysics. In order to solve this problem it is not enough to draw a line of demarcation between them, nor to accept that metaphysics has influence on the evolution of scientific knowledge. What is needed is not the mere acknowledgement of this influence but its explanation in epistemological terms.[26] The axiological controversy concerns the understanding of *truth* either as an autotelic or as an instrumental value. We shall return to these issues later in our discussion. At this point we shall simply state that both Kuhn's and Popper's conceptions emphasize different spheres of scientific activity and treat them accordingly, as the model of the whole. Popper in his methodology does not perceive that scientific theories are formulated not only in order to achieve the truth, but also for their practical utilization. Thus, he provides them no time to prove their usefulness and requires their falsification as soon as possible. Kuhn's vision of normal science considered as a model of science *tout court*, treats science as an instrumentation of historically changing practice, whose modifications are undertaken only in exceptional circumstances. This position disregards the fact that the puzzle-solving activity is not an end in itself through which the scientist proves his ingenuity in the game and gains the recognition of his fellow-scholars. It does not acknowledge that each puzzle is a fragment of a much larger puzzle, whose solution has always been the goal of science.

V

The criticism of falsificationism which was presented in Kuhn's work as well as in the work of Feyerabend and Polanyi, inspired Lakatos to modify Popper's model of the evolution of knowledge. This attempt may be seen as a type of 'counterreformation', that is a manner of neutralizing criticism through the partial acceptance of the theses of critics and the introduction of them into one's own system. But, "if the counterreformation is to be effective, it must be a practical self-criticism."[27] In the case we are considering here, the first target of self-criticism is the obvious discrepancy between the falsificationist model of the growth of knowledge and the very process of evolution of science. Lakatos' proposals are meant to overcome this discrepancy. Let us now see whether these attempts have been successful in this respect.

The falsificationist point of view, as we have presented it above, is modified by Lakatos in two undoubtedly crucial points: (a) first of all, he asserts that falsification consists in the confrontation with experiment not of one theory (together with the background knowledge) but of two competing theories; (b) he rejects the thesis that from the point of view of the evolution of knowledge the only truly interesting result of such a confrontation is falsification.

According to Lakatos, we may speak of the falsification of a theory T_1 only when we have at our disposal a theory T_2, which fulfills the following three conditions: (1) T_2 is of larger empirical content then T_1, so that T_2 has consequences inconceivable within T_1, or even prohibited by it; (2) T_2 explains all the previous successes of T_1, i.e. it contains the whole of the unfalsified content of T_1; (3) at least a part of the new consequences of T_2 has been confirmed by experiment.[28]

The evolution of knowledge is a process of transition from one theory to another and the course of this evolution is determined by specific research programs.

Let us take, according to Lakatos, a series of theories, $T_1, T_2, ...T_n$, in which each subsequent theory results from the acceptance of new assumptions or the semantic reinterpretation of the previous theory. These modifications are to remove the empirical anomalies which the previous theory could not explain. Every successive theory has at least the same empirical content as its predecessor. Such a series of theories, in which each successive one predicts new facts, is called by Lakatos a 'theoretically progressive' series. If, in addition, some of the new consequences have been empirically confirmed, this series is also 'empirically progressive'. A series which does not meet these two requirements, that is one in which every successive theory does not lead to new consequences, or when the new consequences are not confirmed by the experiment, is called a 'degenerating' series.[29] According to the criterion of demarcation, a new theory may be accepted as scientific only if its inclusion into the series will guarantee that the series will be at least theoretically progressive. A theory is considered falsified when it is replaced by a theory T_{n+1} which satisfies the stated conditions.

In respect to classical mechanics, Einstein's theory of relativity was both theoretically and empirically progressive.[30] Galileo's theory that the celestial bodies move in cyclical orbits around the sun may serve, according to Lakatos, as an example of a theory which was not theoretically progressive in relation to its predecessors as it did not prohibit anything that they did not prohibit. Finally, the Bohr-Slater-Kramers theory serves as an example of a theory that is theoretically but not empirically progressive because all of its new empirical conclusions were falsified.

Let us now take a look at the results of Lakatos' modifications.

First of all, the concept of falsification undergoes a profound change. On the basis of falsificationism a theory could be considered as falsified when it was contradicted by an accepted basic statement. On the basis of the modified position described above, it is falsified only when for it is substituted another, competing theory. "There is no falsification before the emergence of a better theory."[31] Therefore, the new conditions for falsification are much stronger: it is not enough for an experiment to contradict a theory, there must be a new theory at hand. Falsification is now equivalent to the demonstration of the falsehood of the old theory, or at least of certain assumptions of the background knowledge.[32]

This conception removes at least partially (why only partially will be demonstrated below) the difficulties which were described in Chapter Two (Section III). We have already noted that the acceptance of a basic statement contradictory to the theory could not be considered as proof of its falsehood, as it cannot be known what is actually falsified, the theory or the background knowledge. The conventionalist solution which separated the theory under test from background knowledge, and made the theory vulnerable to the verdict of experiment, led as we have seen to methodological consequences which did not correspond to the very process of the evolution of knowledge, and which could not be rationally justified. As Lakatos wrote:

> But since this procedure did not offer a suitable guide for a rational reconstruction of the history of science, we may just as well completely rethink our approach. Why aim at falsification at any price? Why not rather impose certain standards on the theoretical adjustments by which one is allowed to save a theory?[33]

According to proposed modification, an experiment which contradicts a theory does not compel us to the instantaneous rejection of that theory until and unless a new competing theory does not emerge. Attempts to save an old theory when it is contradicted by experiments are methodologically as equally justified as the search for a new theory. At the same time, Lakatos provides us with conditions which the proposed modifications must meet: they must guarantee at least the theoretical progressiveness of the series.[34]

It is only with the emergence of a new theory which fulfills these criteria that the defense of the old one becomes methodologically prohibited. But at that point we may say that the experiment which contradicted the old theory was in fact an *experimentum crucis*. Comparing the old theory with the new, we are able to point to their differences and, therefore, discover what was wrong, what in fact was falsified, and how the mistake had been eliminated.

There can be little doubt that such a model comes much closer to the actual activity of scientists who seldom reject a theory without having a better one at their disposal. It also corresponds better to the history of science which demonstrates that sometimes theories persist despite the known facts which contradict them, and sometimes they are rejected without any evidence of anomalies. When a scientist faces the alternative between chaos and imperfect order, as a rule he chooses the latter. This model, according to Lakatos, enables us to eliminate some (but not all) of the conventional elements of falsificationism.[35]

We have already noted that unless we appeal to conventions which serve to separate the tested theory from the unquestionable background knowledge, the experiment alone does not overthrow the theory. Sophisticated falsificationism does not expect an experiment to falsify a theory, but to play the function of an arbiter between competing theories: by falsifying one it confirms the other. Moreover, empirical anomalies are not, according to this model, a necessary condition for the advancement of knowledge. It is possible that they will be discovered only when a new theory with a larger empirical content will be formulated, and some of its new consequences will be judged to be incompatible with the old theory. According to traditional falsificationism, the development of knowledge proceeds through consecutive falsifications of theories which compel us to look for new ones. The existence of a competing theory was considered useful because it could accelerate progress, but it was by no means considered as a necessary requirement. According to Lakatos there can be no falsification of the old theory without a new one. The assessment of anomalies and falsification are two different things. If the strategy of falsificationism can be described as a permanent revolution of experiment against accepted theories, then the strategy of sophisticated falsificationism may be described as the competition of theories, arbitrated by experiment.

If experiment, instead of being a verdict based on presumptive evidence, is to play the role of an arbiter between competing theories, then falsification becomes a relationship between competing theories and their empirical bases. The falsificator of the old theory confirms the new theory; the confirmed consequence of T_2, which could not be deduced from T_1 becomes the falsificator of the old theory. As Lakatos puts it:

Justificationists valued 'confirming' instances of a theory; naive falsificationists stressed 'refuting' instances; for the methodological falsificationists it is the – rather rare – corroborating instances of the *excess* information which are the crucial ones; these receive all the attention.[36]

Let us call attention to the fact that according to Popper's position, as expressed in *The Logic of Scientific Discovery*, a theory is to be accepted not on the basis of confirming procedures but rather by severe and unsuccessful attempts at its overthrow. At this point, however, this position is abandoned by Lakatos. It is difficult to disagree with Agassi that this constitutes a fundamental deviation from the main thesis of falsificationism.[37] Therefore, I believe, that regardless of the appraisal of Lakatos' position, to term it as falsificationism is no more than a *façon de parler*. All that this position has in common with Popper's original thesis, are the claim that scientific statements are theoretical, and the program of constructing the logic of scientific discovery. If the term 'falsificationism' has meant anything specific in the philosophy of science then it has signified that science evolves through successive falsifications of theories exposed to the merciless verdict of experiment, and the postulate of searching for all possible facts which would contradict the accepted theory. Both this thesis and the methodological postulate are rejected by Lakatos.

VI

The modifications of falsificationism presented above result in an important change of the object of methodological appraisal. What now serves as the object of methodological analysis is not a single theory, but a series of theories, being successive steps in the realization of a research program. Such a series is characterized by a certain continuity, which is granted by the research program in the framework of which they were formulated. The history of science becomes a history of research programs rather than of theories.[38]

Such a program is to contain methodological rules indicating the kind of questions that should be avoided, and those which should be posed. To use Lakatos' example, Cartesian metaphysics prohibited the explanation of natural phenomena by means of any hypotheses appealing to forces acting at a distance, and encouraged hypotheses which could reconcile the Cartesian program with such facts as the elliptical orbits of planets which contradicted the Cartesian theory of whirls.

According to Lakatos, every research program may be characterized by its 'hard core', i.e. a set of assumptions which determine the course of research and the problems that requires solution. In respect to these assumptions, inference by *modus tolens* is not allowed: if the results of experiments disconfirm the consequences which follow these assumptions, we are not yet obliged to abandon the program. We must instead make every effort to fortify the hard core of the program with a 'belt' of auxiliary hypotheses which are to prevent the program from falsification. Only to those hypotheses can we apply a *modus tollens*; only they can be directly confronted with experiment. They are to be chosen in a manner which will guarantee that their conjunction with the basic assumptions of the program will survive the confrontation with experiment. A research program is effective when such a procedure leads to a progressive series of theories. Newtonian mechanics provides a good example of such a program: its hard core consists of Newton's three laws. When Newton first formulated this theory, it was swamped by anomalies. However, the followers of the program were gradually able to transform each anomaly into confirming evidence through the introduction of auxiliary hypotheses.

This conception, in Lakatos' opinion, serves to rationalize classical conventionalism. We decide that as long as the introduction of auxiliary hypotheses into the protective belt will widen the empirical content of each successive theory, we will not appraise the program as false even if the auxiliary hypotheses were disconfirmed. In opposition to Poincaré's point of view, Lakatos states that when the program ceases to lead to the formulation of new theories with wider empirical content (that is, when the series becomes degenerating), then the hard core of the program has to be abandoned. This opinion differs also from Duhem's conventionalism, since the rejection of the program does not depend on considerations of its simplicity but on logical and empirical criteria.[39]

Besides the 'negative heuristics' (as Lakatos calls these prohibitions), the research program contains also a 'positive heuristics', that is certain indications and suggestions as to how to change, improve and expand the belt of auxiliary hypotheses. Thanks to them, a scientist may, at least for a time, disregard the discovered anomalies and try to construct hypotheses aiming at their elimination. If the positive heuristics of a progam were as equally well articulated as the negative heuristics, the scientist would have to deal only with mathematical, and not empirical difficulties. Usually, however, this is not the case. Sometimes it happens that the positive heuristics change in the course of the realization of the program, which protects it from degeneration.

Newton formulated his program on the basis of a model consisting of a stationary sun and one moving planet, both bodies being represented by material points. He was quite aware of the fact, however, that he was operating with a simplified model, and that anomalies were, therefore unavoidable. On the basis of this model he succeeded in formulating the inverse law for Kepler's ellipse. This model, however, turned out to be incompatible with his own third law, and he was forced to replace it with another one, according to which both the sun and the planet rotated around their centers of gravity. This change was promoted not by observation, but by theoretical considerations. Then, he constructed a model of a number of planets, abstracting, however, from their mutual gravitational influence. Now, in turn, he was forced to consider a model according to which the celestial bodies were no longer material points, but solid bodies. As is well known, the mathematical solution of this problem took the next ten years of his life and delayed the publication of his work. Finally, he constructed a model which took the interplanetary gravitation forces into account and worked on the problem of perturbations. Lakatos believes that the fact that Newton progressed from one model to another while fully aware that all of them were provisional, and even knowing what was wrong with them, provides the best evidence for the existence of 'positive heuristics' within a program.[40]

VII

I have already noted that Lakatos' conception undoubtedly arose under the influence of the criticism of falsificationism, and that it constituted an attempt at neutralizing this criticism by a partial acceptance of its theses and by their inclusion into his own theory.

And so the concept of a research program, as formulated by Lakatos, undoubtedly corresponds to the concept of the paradigm. Although Lakatos rejects Kuhn's concept of the monoparadigmatic evolution of science and claims that in science we always deal with various competing research programs, nevertheless, what he calls the development of a program strictly corresponds to the articulation of a paradigm within the ramifications of normal scientific activity.

In turn the concepts of negative and positive heuristics correspond to Kuhn's thesis concerning the 'rules of the game' contained within the paradigm, or concerning the rules of solving puzzles which indicate to scientists what questions they should ask and in which direction they should look for their solutions.

The concept of a protective belt of hypotheses and the rule prohibiting the application of *modus tollens* in respect to the 'hard core' of the program is simply a translation into the language of methodological conventions of Kuhn's statement that scientists in their normal activity usually try not to falsify the accepted paradigm, but to solve the puzzles which it has provided and which, so far, have resisted any solution. Lakatos' acceptance of this rule is also in agreement with Kuhn's thesis, that any failure on the part of an investigator in solving such problems, and the unsuitability of a hypothesis he has proposed, are not usually taken as evidence, that the program was wrong but that the investigator was simply not ingenious enough.

The notions of progressive and degenerating series of theories in turn correspond to Kuhn's concept of the development of normal science and of crises which lead to the emergence of a new paradigm (research program).

We should not, however, be misled by these convergences between the two presented points of view. They are a result, as we have already said, of a tendency to connect the logic of scientific discovery more closely with the real facts of the evolution of science, and to meet the requirement that the theory of knowledge should reflect the phenomena uncovered by history of science. These convergencies do not mean that Lakatos has accepted Kuhn's model of the evolution of science. The 'neutralization of criticism' is not an acceptance of its theses, but an attempt at assimilating them in a way which provides the possibility of defending his own position in basic matters and to make it more resistant to criticism. More precisely: if according to Kuhn, the transition from one paradigm to another cannot be in any way explained only in logical or methodological terms (a paradigm is like a prison which we cannot leave by our own efforts), then, according to Lakatos, his methodology of research programs explains this transition and, therefore, it is absolutely unnecessary to go beyond logic and methodology. This is what I had in mind when I termed the modifications introduced by Lakatos into the falsificationist position as a 'counter-reformation'. The fact that he accepted certain theses of the critics of falsificationism is by no means equal to the convergence of these positions. These two visions of the evolution of science remain as far apart from each other as they ever were.

The effectiveness of the proposed methodological modifications depends then on the answer to the following questions: in what measure is the modified model of the logic of scientific discovery (or methodology of research programs) free from the difficulties which had previously been pointed out by its critics? In what measure is it capable of accounting for the actual process of development of knowledge? Is it really immune to the criticism which was directed at the falsificationist position, and was considered so relevant that it led to the modification of that position? In brief, to what degree was the self-criticism effective?

Before we proceed with the answer, one short remark is necessary. It appears that we should differentiate two different theses of Kuhn's which were attacked by his critics, but which are not connected with each other in such a manner that by accepting one we would be forced to accept the other. The first concerns the monopoly of a paradigm in the period between revolutions; the second claims that the transition from one to another paradigm cannot be explained exclusively in methodological terms. The former, as we have already noted, has not been accepted by any other investigators and we will not repeat their arguments here. The latter, on the other hand, concerns a problem which is crucial in the context we are dealing with, as the modifications introduced by Lakatos into falsificationism were to demonstrate that it was groundless. In any case there is no reason to believe that the rejection of the conception of a monoparadigmatic evolution of science in the periods between revolutions provides a solution for the second, more fundamental problem.[41]

The methodological criteria formulated by Lakatos were concerned, as we have seen, with two essential problems; first of all, the transition from one theory to the next *in the framework of a research program*, and secondly, *the transition from one program to another*.

The first problem, as I see it, does not seem very controversial. The conception of the falsification of a theory as a result of the acceptance of another one abolishes the arguments advanced against falsificationism in this matter. What remains questionable is Lakatos' claim in accepting this conception of falsification that there is no need of a convention stating that the background knowledge is unquestionable. This would indeed be the case if two competing theories shared the same background knowledge, that is, if the two theories were connected with exactly the same ontological and epistemological assumptions as well as with the same kind of measuring procedures. Only in such a case could the result of an experiment, falsifying one theory and confirming the other, unequivocally indicate where the mistake lies. For it would be obvious then that it is the falsified theory which is responsible for the mistake and not the background knowledge. In the case, however, where the background knowledge of the two competing theories is not identical, the verdict of an experiment cannot be conclusive. For it would be impossible to discern

whether the falsification of one theory and the confirmation of the other
results from the difference in their content or from the difference in the
underlying assumptions. It is by no means clear that competing theories
satisfy this condition, even within the framework of the same research
program, (unless the concept of a program would be understood in such
a manner that it would encompass the whole of the background knowledge, and that any change in the background knowledge constituted, by
definition, a change of the program. It does not, however, seem apparent
that this is how Lakatos understands the concept of a research program).
Therefore, the claim that Lakatos' position 'weakens' the conventionalist
element of falsificationism is by no means evident. It is unfortunate that
the concept of a research program with which Lakatos operates is not
unequivocal enough so that it might provide the solution to this problem.
This matter, however, is not of great importance, as Lakatos himself
stated that the conventionalist element cannot be completely eliminated.

Important doubts arise, however, in respect to the problem of the transition from one paradigm (research program) to the next, that is in respect
to the methodological criteria which are to decide whether the old program has degenerated to the point that it must be replaced by the new one
or if it still provides any hope for future successes.

Let us suppose that in attempting to solve a particular problem, which
is posed by the program, we encounter an empirical anomaly and, according to the prohibition of applying *modus tollens* to the basic assumptions of the program, we try to reconstruct the protective belt of hypotheses. As a result we formulate a new theory T_{n+1}, but now it appears that
its introduction into the series leads to the degeneration of the program.
The resulting series is not progressive, because the new theory either does
not lead to any new consequences, or the resulting consequences cannot
be empirically confirmed. The old theory T_n has not been falsified, and
the new one has not been accepted. However, according to the negative
heuristic of the program (the prohibition of applying *modus tollens*), we
should not infer that the basic assumptions of the program are false.
Therefore, we find ourselves back in the same place as we were before
the formulation of the theory T_{n+1}. What, then, are the methodological
rules which would permit us to decide whether we should repeat the procedure in the hope that next time we will find a theory T'_{n+1} which would

meet the conditions of progressiveness or abandon the program and look for another one?

This is, of course, the same difficulty with which the falsificationist model had to deal on the level of theories, before Lakatos proposed his modifications. By introducing the concept of the research program he simply elevated this difficulty to a higher level; by solving the problem of choice between theories through the introduction of the concept of the program, he left open the problem of choice between programs.

The solution that we should not abandon the old program prior to the emergence of a new one, that there can be no falsification of a program exclusively on the basis of experiment, i.e. the same solution which he proposed for the problem of the falsification of theories, is unsatisfactory in a number of respects.

First of all, if all scientists were to proceed according to this postulate, a new research program would never emerge. The rule proposed by Lakatos might be applied only if someone had violated it. The necessary condition for the proper methodological procedure is a violation of its rules by someone who was in search of a new program prior to the falsification of the old one. Popper's ethically attractive proposal to undertake a maximal theoretical risk turns here into its own contradiction. This results from the fact that in the case of an experiment which falsifies a theory, Lakatos' rules specifically state: the theory is wrong, seek another one, but do not abandon the old one until you have found a better one. An experiment cannot, however, help us evaluate a research program since the negative heuristic guards its basic assumptions against falsification. On the basis of Lakatos' rules, the attempts to save the old program are as equally justified as a rejection of the program and the search for a new one. No methodological rules can, of course, decide the question of how long the attempts to save the old program remain rational.

Second, let us assume that we will disregard the problem of from where and how did the new competing program originate. We may assume, in contradiction to Kuhn, that in science no program ever has a full monopoly, that there are always several programs being developed by various schools. (As a matter of fact, it is my conviction that the thesis that there are *always* several research programs, is just as false as the concept of a monoparadigmatic development of science in the period between revolutions. In addition, the correctness of this thesis is dependent on how broad-

ly – or how narrowly – one understands the concept of a research program.) In order that Lakatos' methodological postulates may resolve our
problem, it is not enough always to have several competing programs.
What is necessary is a fulfillment of a stronger requirement: one of the
competing programs must provide a solution to the discovered anomaly,
and at the same time it should not give rise to problems, unencountered
by the old programs. It goes without saying that as a rule the situation is
not always so simple. The history of the debate between the corpuscular
and the wave theory of light may serve here as an excellent example not
only of the fact that there are usually several research programs and that
a return to previously abandoned programs may be fruitful, but also of
the fact that when there are two competing theories, it is not necessarily
always the case that one of them will provide a progressive and the other
a degenerating series of theories. Prior to the work of de Broglie both
programs were degenerating. The wave theory could explain diffraction,
but could not deal with the problem of the photoelectric effect; the corpuscular theory, on the other hand, could effectively explain the photoelectric effect, but the phenomena of diffraction constituted an anomaly
for it. The real crisis in science occurs when, using Lakatos' terminology,
no program can provide a progressive series. Therefore, even if we justi-
fiably reject Kuhn's monopolistic thesis that in periods between revolutions
we are dealing only with one program, our problem is by no means solved,
and the argument presented above still stands.

Finally, what in reality are these research programs which Lakatos
talks about? We are forced to return to the same questions which were
raised in Section III in connection with Kuhn's concept of the paradigm.
In addition, however, there emerges one additional question.

We have already quoted Lakatos' example of Cartesian metaphysics
as the source of the programmatic assumptions of Cartesian physics.
Considering the position which he appears to be defending, this is a perplexing example. Is it not a resignation from the whole conception of the
demarcation of science from metaphysics, that is from the very goal of
falsificationism? To be sure, Popper never claimed that metaphysics is
meaningless, but he always sought to exclude it from science. Here,
however, it seems that metaphysics must be somehow let in, by the back
door so to speak, as it is recognized that certain metaphysical assumptions lie at the root of a program. Are we, therefore, correct in concluding

from this that Lakatos generally abandons the falsificationist (and for that matter any) solution to the problem of demarcation? Unfortunately, what he has written on the subject of research programs is not clear enough to allow an answer to this question. Nowhere did he explicitly say whether statements which do not fulfill the criterion of demarcation can belong to the core of the program. The prohibition of applying *modus tollens* would indicate that they are to be falsifiable, as otherwise this prohibition would not be necessary at all. As a result of this it would appear that they are not metaphysical in the falsificationist understanding of the term. The example quoted above, however, appears to indicate quite the opposite.

In conclusion, it should be noted that Lakatos' 'self-criticism' is not fully effective as it does not solve the crucial question of the progression from one program to another. The problem of whether this progression could be explained in purely methodological terms, which is what was to be proved, remains open. By the same token, the thesis that the logic of scientific discovery[42] may constitute a satisfactory base for the reconstruction of the process of the evolution of knowledge remains unjustified.[43]

As a result of the confrontation of the positions which we have just presented, we are faced with a number of problems. First of all, the very concept of a scientific revolution remains unclear; second, we must seek to explain the role of factors which cause the transformations of theoretical knowledge and try to elucidate the character of these transformations. Finally, we must investigate the problem of the continuity of the evolution of knowledge; that is, the problem of the relationship between new and old theories. We shall now undertake the discussion of these questions.

NOTES

[1] Cf. M. Polanyi, *Personal Knowledge*, London 1958, and his *Knowing and Being*, especially Part III, London 1969.
[2] T. S. Kuhn, *The Structure of Scientific Revolutions*, p. 121.
[3] *Ibid.* p. 3.
[4] In his 'Reply to Criticism' (in *Criticism and Growth of Knowledge*) as well as in the 'Postscript' to the second edition of *The Structure of Scientific Revolutions*, Kuhn explicitly refers to the conception of untranslatable languages. Feyerabend defends an even more radical thesis; he speaks of the complete incommensurability of successive theories. (See 'Consolation for a Specialist' in *Criticism and the Growth of Knowledge*, pp. 59–91, especially pp. 61–65).
[5] M. Masterman, 'The Nature of the Paradigm', in *Criticism and the Growth of Knowledge*, pp. 59–91 (cf. especially pp. 61–65).

[6] See the 'Postscript – 1969' to the second edition of *The Structure of Scientific Revolutions*, p. 176.

[7] T. S. Kuhn, 'Logic of Discovery or Psychology of Research', in *Criticism and the Growth of Knowledge*, p. 7.

[8] The papers of J. W. N. Watkins, S. E. Toulmin, L. P. Williams, K. R. Popper, P. K. Feyerabend, and I. Lakatos in *Criticism and Growth of Knowledge*, Cambridge 1970.

[9] It seems that Kuhn's conception of the monoparadigmatic evolution of science resulted in part from his studies of the history of astronomy, where Ptolemy's theory had a monopolistic position for centuries.

[10] K. R. Popper, 'Normal Science and its Dangers', in *Criticism and the Growth of Knowledge*, p. 53.

[11] See *The Structure of Scientific Revolutions*, pp. 57–58.

[12] See my 'Introduction' to the Polish edition of *The Structure of Scientific Revolutions*.

[13] Cf. W. Heisenberg, *Physics and Philosophy*, New York 1958.

[14] See the 'Postscript – 1969' to the second edition of *The Structure of Scientific Revolutions*, pp. 174–176.

[15] T. S. Kuhn, 'Logic of Scientific Discovery or Psychology of Research', in *Criticism and the Growth of Knowledge*, p. 7; see also *The Structure of Scientific Revolutions*, Chapters 3 and 4.

[16] Feyerabend is right when he remarks that Kuhn does not analyze this problem at all (cf. "Consolation for the Specialist' in *Criticism and the Growth of Knowledge*, p. 210).

[17] *The Structure of Scientific Revolutions*, pp. 35–36.

[18] *Ibid.*

[19] *Ibid.* pp. 35–36.

[20] *Ibid.*, pp. 36–38.

[21] *Ibid.*, pp. 170–173.

[22] *Ibid.*, p. 170.

[23] *Ibid.*, p. 35.

[24] Cf. H. Hall, 'Scientists and Politicians', in Barber and Hirsch (eds.), *Sociology of Science*, Glencoe 1962, p. 275.

[25] *The Structure of Scientific Revolutions*.

[26] Cf. M. Wartofsky, 'Metaphysics as Heuristic for Science', in *Boston Studies in the Philosophy of Science*, Vol. III, Dordrecht 1967, pp. 123–172.

[27] L. Kołakowski, *Notatki o współczesnej kontrreformacji* (Notes on Contemporary Counterreformation) Warsaw 1962, p. 3.

[28] I. Lakatos, 'Falsification and the Methodology of Scientific Research Programmes', in *Criticism and the Growth of Knowledge*, p. 118.

[29] *Ibid.*, pp. 118–121.

[30] This example seems odd because it assumes that both theories were successive steps in the process of realization of the same research program.

[31] I. Lakatos, *op. cit.*, p. 119.

[32] This corresponds to the opinion of Feyerabend who says that the best criticism of a theory is provided by a new theory which replaces it. ('Reply to Criticism', in *Boston Studies in the Philosophy of Science*, Vol. II (ed. R. S. Cohen and M. W. Wartofsky), Dordrecht 1965, p. 227.

[33] I. Lakatos, *op. cit.*, p. 117.

[34] Strictly speaking, the theory resulting from modifications which are to save it from contradictions with experiments, is already a new theory, a successive component of the series.

[35] We have still to decide by convention which basic statements are testable, and which do not need any further justification. And only by convention can we decide whether the obtained statistical distribution does or does not contradict a statistical theory. (Cf. *ibid.*, pp. 125–131).

[36] *Ibid.*

[37] See above, Chapters 4 and 5.

[38] I. Lakatos, *op. cit.*, p. 132.

[39] *Ibid.*, p. 134.

[40] *Ibid.*, pp. 135–136.

[41] Lakatos refutes both of these theses. Feyerabend, who criticises the conception of the monoparadigmatic evolution of science believes, however, as Kuhn does, that the transition from the old to the new paradigm cannot be explained only in logical terms.

[42] As I have often mentioned above, Popper, and following him Lakatos, term their theories the *logic* of scientific discovery. There may be some degree of doubt, however, as to whether there is any justification for naming these conceptions in such a manner. The name, of course, is not of primary importance, but it may lead to certain confusions, especially when it is connected with certain expectancies and premonitions. Thus we could, for example, expect that a theory so named offers a pattern of logical transformations which science undergoes in the course of its evolution, i.e. a pattern of certain syntactical relationships between the statements of the language of the evolving knowledge as well as semantic relationships connecting them with reality. In this case, however, this is not the case. The evaluation of scientific statements is to proceed on the basis of their features as known from logic, and the process of the evolution of knowledge is presented as the result of the systematic application of these methods of control and evaluation. Therefore, we are not dealing here with anything which could be termed a 'diachronic logic', or the logic of evolution in the strict sense of the term, but with the conviction that the evolution of knowledge proceeds (disregarding accidental deviations) as a result of the application of the indicated methodological criteria. It seems that the emphasis on this difference is not without significance, especially as we know of alternative conceptions which aim at the construction of the logic of the development of science in the sense indicated above. What I have in mind here are the unending unsuccessful attempts (since Hegel) at the construction of a so-called dialectical logic, as well as the conception of the explanation of the evolution of knowledge in terms of formal logic, as represented by R. Suszko. (See R. Suszko, 'Logika formalna a niektóre zagadnienia teorii poznania' [Formal Logic and Certain Problems of Epistemology] in *Logiczna teoria nauki* [*The Logical Theory of Science*], Warsaw 1966, pp. 505–578 and 'Formal Logic and the Development of Knowledge' in *Problems in the Philosophy of Science* (ed. by I. Lakatos and A. Musgrave), Vol. III, Amsterdam 1968, pp. 210–222 and the following discussion on pages 223–230.)

[43] After reading these remarks, Lakatos told me in a private conversation that as many other of his critics, I have not understood his position. He stated that his methodology was not intended at all to solve the problem of how the transition from one program to another really occurs or how scientists should proceed in such situations in order to be rational, but that it had to provide a basis for a rational appraisal of theories *ex post*. I will only say here that if that were the case, it would be hard to understand the nature of his disagreement with Kuhn's thesis that the actual process of changes in the content of knowledge cannot be accounted only in terms of methodology, and why Kuhn's position is to be qualified as irrational if his own were a rational one.

CHAPTER VII

REVOLUTIONS IN SCIENCE:
THE ACCUMULATION OF KNOWLEDGE AND
THE CORRESPONDENCE OF THEORIES

I

It will be convenient here to begin our analysis of the character of revolutions in science, that is of the mutual relationship between pre- and post-revolutionary states of knowledge by presenting Roman Suszko's conception of diachronic logic. It must be stated at the outset that this conception does not aim at the reconstruction of the history of science nor is it supposed to account for the actual procedure of scientists. It only seeks to explain, by means of the time-tested tools of logic, the relationships between successive formulations of scientific theories. This conception requires a solid logical analysis and discussion, for which I do not feel sufficiently competent. However, it will be of interest to us here insofar as it allows us to distinguish between various types of transformations which occur in the process of the evolution of knowledge and at the same time it indicates which types of transformations cannot be accounted for exclusively in logical terms.

When we seek to analyze human knowledge from the point of view of formal logic, states Suszko,[1] we are dealing with an epistemological relationship: subject – object [S.M.]. The main component of the subject is the formalized language L, while of the object –, a model M of that language. (The term 'model' is understood as in semantics.) There are specific semantic relationships between the language and its model, such as those of denotation, truth, falsehood. The language contains a set of statements T which are currently accepted by the subject. The set T does not exhaust all the sentences which may be formulated in the language L. This means that not every sentence a, or its negation, of the language L belongs to T. Furthermore, not only sentences which are true in relationship to M belong to the set T, as the subject may commit mistakes and accept false sentences. (The terms 'truth' and 'falsehood' are understood here in Tarski's sense.) The set T encompasses also a certain subset of statements A consisting of analytic sentences, or more precisely, of sentences which the subject interprets as analytic. Therefore, the system (L, A, T) M is somewhat like a semantic system and may be an object of logical analysis.

In contrast to synchronic logic, which studies the structure of such systems, diachronic logic is to study the transformations which they undergo in the process of the evolution of knowledge. It is, therefore, interested in the transformations of epistemological relationships, that is in the transition from a system (L, A, T) M to a system (L^x, A^x, T^x) M^x. Such transformations may be considered as successive steps in the development of knowledge.

The first, and simplest, case analyzed by Suszko occurs when the object of study does not undergo any change, i.e, when $M = M^x$. It is obvious that in such a case $L = L^x$. This transformation then can be formalized as (L, A, T) $M \rightarrow (L, A^x, T^x)$ M. This type of transformation is typical for the evolutionary process of the broadening of knowledge about a given object. They may describe the process of the expansion of the set T through the inclusion of new statements of the language L, which may be either true or false, and account for the elimination of certain statements previously accepted, but now proven to be false. In this case the set of axiomatic statements A does not undergo any change $(A = A^x)$ or may undergo only specific changes, namely satisfying the condition that no sentence previously considered as axiomatic ceases to be considered analytically true, although it may no longer be considered as an axiom. This means that the manner of axiomatisation of a theory may, but does not necessarily have to, undergo a change.

Suszko states that he is unable to specify any formal characteristic of the transformation T/T^x, which would characterize the process of the evolution of knowledge. But he considers that is is possible to say something about a series of such transformations. Thus, it may be assumed that such a series converges to a limit $T_n = \mathrm{Ver}(M)$, that is to the state when all the sentences true in respect to M will belong to the set T of statements accepted by the subject. This would signify that every false sentence of the language L would belong to a small number of sets T, but that every true sentence would belong to almost all the sets in this infinite series. In other terms, each false sentence would be eliminated following a finite number of procedures while every true statement would remain forever in the series. Suszko admits "that we cannot select any convergent sequence of transformations as above and command the subject to behave in this prescribed way",[2] as this would be equivalent to providing an algorithm for achieving absolute truth about the object under investigation. Regardless, however, of which strategy could be suggested by methodology, and whether it could in fact provide any, the case described here strictly corresponds to the cumulative character of the evolution of knowledge. The object does not change, the meaning of basic notions remains unchanged (L=const), truths once achieved are handed down, mistakes are successively eliminated and the set T is constantly being enriched by new true statements. There remains only the possibility of the reorganization of achieved knowledge through the transformation of its axiomatic base.

In distinction with evolutionary change, when the language and model remain constant, in the case of revolutionary change we are not dealing with the convergence of series of transformations, and the principle of the permanence of axioms becomes restricted. Suszko distinguishes two such cases: of weak and strong scientific revolutions.

In the case when $M \neq M^x$ "it is clear that there is only one reasonable general assumption which can be made concerning the change of the object within the development of knowledge. It is the assumption that the former object M is *a sub-model* of the later model M^x. We can say also that the model M^x is *an extension* of the model M and we write: $M \subset M^x$.[3] This signifies that the new model contains new objects and new relationships which have no counterparts in the old language L, and, therefore, we are dealing with a new language L^x, which is an extension of L. The transformation now has the form (L, A, T) $M \to (L^x, A^x, T^x)$ M^x. At this point not every sentence which is true in the model M will also be true in M^x, and *vice versa*. In this case as well, it is impossible to indicate any formal characteristics of such transformations.[4] The author states, however, that if M is a sub-model of M^x, then the axiomatic basis will be preserved at least to some degree. He proceeds to analyze two such cases, when $M \subset M^x$.

In the first case, the denotations of the old language L do not undergo any change. The old model is simply a constituent part of the new one and the only difference is that L^x contains certain new constants which were not contained within L. This means that new concepts are added to those with which the subject already operates, while the old ones are not elimi-

nated. In this case the axioms which were true within the old model, remain true in the new one. Thus, here too, we are dealing with a process of
evolution – with the difference, that if in the previous example the development of knowledge was based exclusively on the change of the set of
statements accepted by the subject which could be expressed in the
language L, then in the present case this change also encompasses the
appearance of new concepts. Therefore, L^x is richer than L. This type of
modification "...may be seen as *a weak revolution*".[5]

If this understanding of Suszko's idea is correct, he would accept as an
example of a 'weak revolution' the introduction of the concept of the spin
into quantum mechanics. If so, then the notion of a 'weak revolution'
would be encompassed by what Kuhn calls an 'articulation of the paradigm', and Lakatos terms 'the modification of the protecting belt of
auxiliary hypotheses'. The meaning of concepts of the language L would
not undergo any change, and therefore all observational statements would
retain their meaning, provided that while language L could not account
for certain empirical facts because of a lack of appropriate concepts,
language L^x, as a result of its enrichment, could account for them. Thus,
while in the case of evolutionary changes the language L is enriched only
by new statements, in the case of a weak revolution it is enriched addition-
ally by new concepts. This kind of change may be called a revolution if we
accept, as Suszko does, that any introduction of a new concept into the
language results in its modification.

In the second case, which Suszko terms a 'strong revolution', the change
consists in the appearance, among the constants of the language L^x of a
constant such that all the statements of the language L, if they are to be
true in M^x, must be relativised in respect to this constant. This means
that, "if α is a sentence of the former language L true in the former object
M then (as may be proved) the sentence α_ϕ which is the result of relativisation of bound variables in α is true in the later object M^x".[6] Thus, the
truth-value of certain sentences changes, and "the subject assumes only
such new axioms as allow him to prove the old axioms when they are
restricted to the old universe of discourse".[7] The revolutionary change
means that certain old axioms are also rejected.

'Strong revolutions' would then denote such changes in the history of
science as the transition from Newtonian mechanics to the theory of relativity, or from classical physics to quantum mechanics. The thesis that

every statement which is true in M would be true in M^x after its relativisation, means that the old theory is a special case of the new. In the case of the transition to quantum mechanics the constant in relationship to which this relativisation was performed would be Planck's constant; in the case of the transition to the special theory of relativity, the ratio of the velocity of a moving object to the velocity of light.

The question arises, to what degree do these three cases exhaust the transformations occurring in the process of the evolution of knowledge?

If the most radical changes in science could be described in this manner, it would mean that (a) the evolution of knowledge is a *cumulative* process since every successive theory would explain all the facts explained by the old theory; and (b) that between the old and new theories there exists such a relationship that the new universe is always an extension of the old one $(M \subset M^x)$.

By stating that revolutionary change, weak or strong, consists in transformations of the object of study M, Suszko immediately assumes that the new model is always an extension of the old one. It is just this assumption which seems questionable. Its acceptance, however, appears to be unavoidable on the grounds of his conception. For, without assuming such a relationship between M and M^x, and, consequently, between L and L^x, the attempt to describe all the change in science in logical terms would be unrealizable. Without fulfilling this basic condition, diachronic logic cannot describe all the possible types of transformations.

The acceptance of this assumption, however, seems questionable for the following reasons:

First of all, we know of certain transformations in the history of science,[8] when M^x does not contain certain objects and relationships present in M. As examples we note the refutations of the theories of phlogiston, caloric, ether, or of action at a distance. In all these cases, although it is true that $M \neq M^x$, it is not true that $M \subset M^x$. In answer to this remark, Suszko argues that "we may always assume that the old language is a sub-language of the new one. In particular, we may suppose that the expression 'phlogiston' belongs to contemporary scientific language, that its denotation is the void set, and that the nonexistence of phlogiston is lastingly asserted."[9] It seems, however, that such an argument is not only artificial, but also that it cannot be reconciled with the conviction that the language of a theory is formalized, as in such a case there would be no such term, or statement, which could not be treated as belonging to the language of science.

Secondly, and this seems to be the essential point, the thesis that sentences accepted in the language L are also true after their appropriate relativisation in L^x, by no means signifies that they will have the same meaning in both languages (and in both models). This would only be the case if the terms of the language L had the same interpretation in both models, M and M^x. (I obviously have an empirical interpretation in mind here.) The statement 'the mass of a body does not change as a result of its movement' remains true after its 'appropriate relativisation' in the theory of relativity but this does not mean that the empirical meaning of such terms as mass, movement, space and time remains unchanged in the transition from M to M^x. This would only be so if at least some of the statements of L and L^x were purely observational and if their meaning were totally independent of the theoretical context in which they appear, that is of the theory within which they operate, and of the background knowledge which accompanies this theory. In other words, it seems that Suszko's thesis, that M^x is always an extension of M, would be true only if it were possible to reconstruct the language of the empirical sciences on the basis of empirical and axiomatic semantic rules and to present the evolution of knowledge as a sequence of transformations of such a language.

II

From the above remarks, it follows that when we undertake to study the relationship between the state of knowledge before and after the change which is termed a revolution, we must deal with the following problems.

First of all, can the new theory explain all the *natural phenomena* which the old theory was able to explain, in other words, does the *accumulation of knowledge* really occur?

Second, whether the old and the new theory are connected by the relation of *correspondence*?

In this context, we must draw attention to the fact that the thesis of correspondence may be understood in two ways: (a) in a weak manner, according to which the old theory appears as a particular case of the new one in a formal sense that is independent of the empirical interpretation of its statements; (b) in a stronger sense, when the statements of the old theory, after their relativisation, are not only true in the new one, but moreover they preserve their empirical meaning.

In the first instance, we may speak of a formal or syntactic correspondence, while in the second one, of empirical or semantic correspondence.

Let us note that while a negative answer to the question of accumulation would also imply a negative answer to the question of correspondence (especially in its stronger interpretation), it is not the case that the opposite is also true. A new theory may explain all the natural phenomena that were explained by its predecessor, but this does not imply that the theories are linked by the relationship of correspondence.

The distinction between these two questions, dealing respectively with accumulation and correspondence, is a result of the position we defended in Chapter Four, according to which scientific facts are not bare empirical facts, but interpretations of natural phenomena. Therefore, the very natural phenomena may, but do not have to, constitute different scientific facts within successive theoretical frameworks. Only in the case when, following the approach of radical empiricism, we fail to notice the impact of the cognitive apparatus upon the understanding of scientific facts, do the problems of the accumulation of knowledge and of the correspondence of theories merge into one. And in such a case, the negation of correspondence between theories implies the rejection of the thesis of the cumulative character of the evolution of knowledge. What results is the confusion of these facts: that in the process of evolution of knowledge, the amount of our information about the universe constantly increases, and that this information is ordered by successive theories in different ways, changing our world perspective.

To take one example, man has known since ancient times that heavy bodies fall and that while they fall their velocity increases. However, this well-known phenomenon was by no means interpreted as the same scientific fact in the frameworks of Aristotelian physics and of Galileo's mechanics.[10] Aristotle and his medieval followers, regarding space as a finite and hierarchically ordered physical continuum, containing certain specific directions and places where infinite rectilinear movement is impossible, saw this phenomenon as the aiming of a body towards its natural place. Every case of free movement of a body, whether upwards or downwards, was understood as a case of teleologically caused movements. For

Galileo, on the other hand, who understood all physical processes as occurring not in an anisotropic, finite physical space, but in an isotropic, infinite geometrical space, in which no directions or places can be discriminated, free fall of a body would not be treated in the same manner. And for this reason Galileo studied a *totally different fact* than did his predecessors, although, quite obviously, they were all dealing with the same natural phenomenon. This is of crucial importance, for the manner in which one studies the facts and for the kind of questions that one raises.

If one assumes that any free motion is 'aiming at something', then he will be unable to discern the essential difference between the formulation that velocity increases as a body approaches its destination and the statement that it increases as the body moves away from its starting point. Aristotelian physics favored the former solution. Tartaglia, however, one of the predecessors of Galileo and a partisan of the theory of impetus, considered both formulations to be equivalent. But for Galileo, and for Benedetti, his teacher of physics at Padua, these two statements could not by any means be considered equivalent. For them the first solution is simply nonsensical, for it is connected with the false concept of movement towards a goal: velocity cannot increase as a body approaches its destination as there is simply no such place.

This difference in the conceptual apparatus which we have described above had one other important consequence. For, on the basis of Aristotelian physics, natural infinite (rectilinear) motion that is not aiming at any natural place could not in any manner constitute a physical fact, as such a movement was simply impossible because of the finiteness of space. From Galileo's point of view, however, such movement was not only possible, but it demanded serious attention and analysis. And it was the analysis of this kind of movement which led to the formulation of the law of inertia.

Irrespective of how much time and energy Galileo might have spent observing falling stones, he could have formulated neither the law of free fall nor the law of inertia if he had not previously freed himself from Aristotelian cosmology and the associated concept of space. As A. Koyré has convincingly demonstrated, the new physics was born out of the geometrization of the concept of space, or more precisely, from the steady erosion of Aristotelian cosmology. And it is as a result of this process that the same phenomenon of the fall of heavy bodies, which had been observed for centuries, became a totally new scientific fact, different from what it had been for Aristotle or for the partisans of the medieval theory of impetus such as Leonardo, Tartaglia, Benedetti and many others. And, we might add, this same phenomenon of inertial motion became a new scientific fact within the ramifications of Newtonian physics through the introduction of the concept of gravitation, and subsequently within the realm of the general theory of relativity which introduced the concept of curved space.

There is something in the real structure of the universe which connects all these scientific facts, some common factor which may appear in different forms in successive theories, but which no scientific theory can ignore. And in this sense, the evolution of knowledge is undoubtedly a cumulative process. But, as it is not difficult to notice, such an understanding of accumulation by no means implies that successive theories must correspond to one another, nor that each successive model is an extension of its predecessor. If in one theory the free fall is treated as the aiming of a body at its natural place, in the next as a motion along a cyclic orbit in an infinite, isotropic geometrical space and in the third as motion along a geodetic in the finite but unlimited Riemannian space, then it seems impossible to speak of the (semantic) correspondence of theories explaining this phenomenon.

III

As we have already noted, the thesis of correspondence between successive theories is logically stronger than the thesis of the accumulation of knowledge. For it states not only that phenomena which are explained on the basis of theory T_n must also be explainable by T_{n+1}, but it also states that these explanations must be linked by the relationship of correspondence.

In order to ascertain to what degree the fulfillment of the principle of correspondence may characterize changes occurring in our knowledge, let us investigate its weaker (formal) variant.

As our example, we have chosen the three successive formulations of Ohm's law,[11] which expresses the dependent relationship among the difference in voltage $(V_2 - V_1)$, the current (I), and the resistance (R) of a conductor: $V_2 - V_1 = IR$.

Assuming that the general form of a scientific law can be presented as a formal implication $\prod_x W(x) \to Z(x)$ we shall, following W. Mejbaum, term the expression $W(x)$ as the conditions of relevance, and the set of objects fulfilling these conditions as the range of relevance of the law. By the term $Z(x)$, we understand the physical dependence stated by the law.

The original formulation of Ohm's law stated that:

(1)   for all x, if x is a homogeneous conductor of electricity there exists a relation, $V_2 - V_1 = IR$.

Thus the range of relevance embraced all homogeneous conductors, while the condition of relevance was that the voltage, the current, and the resistance do not change with time.

At the moment when it was realized that the law is not fully exact, as it did not take into account such a factor as self-induction, the formulation had to be changed and replaced by $RI + L(dI/dt) = E$, where L is the coefficient of self-induction, dI/dT the differential of current in time, and E the electromotive force. This modification could be expressed in two ways:

(2)   for all x, if x is a homogeneous conductor without self-induction, then Ohm's law is fulfilled;

(2')   for all x, if x is a homogeneous conductor, the relation $RI + L(dI/dT) = E$ occurs.

When, in turn, it was discovered that this formula did not describe the process precisely as it did not take into account the capacity of the circuit, the equation $RI + L(dI/dt) = E$ was replaced by the equation $RI + L(dI/dt) + Q/C = E$ where Q means the electrical charge, and C stands for capacity.

As a result it was possible to state that:

(3)   for all x, if x is a homogeneous conductor without self-induction and of infinite capacity, then Ohm's law is fulfilled;

(3')   for all x, if x is a homogeneous conductor of infinite capacity, then $RI + L(dI/dt) = E$;

(3")   for all x, if x is a homogeneous conductor, then $RI + L(dI/dt) + Q/C = E$.

What, then, constitutes the difference between these formulations?

In comparing (1), (2) and (3), we must conclude that they describe the same physical relationship, Ohm's law, and that they differ only with respect to the conditions of relevance. Successively, the range of relevance is narrowed down. By analogy, the same can be said of (2') and (3'); here the transformation results in the limitation of the range of relevance of the equation $RI = L(dI/dt) = E$.

By comparing, on the other hand, the formulations (1), (2'), and (3''), we may conclude that the range of relevance remains unchanged, but that they differ in their formulations of a physical relation.

If we assume then, following Mejbaum, that "two formulations express the same physical law if and only if they do not differ in any respect, except for the range of relevance",[12] then we are forced to conclude that the formulations (1), (2) and (3) all express Ohm's law, while the formulations (1), (2') and (3'') express different laws.

It is immediately apparent, however, that the formulations (2) and (2') and by analogy (3), (3') and (3'') differ from each other only formally. The changes in our knowledge may be described here either as modifications of the range of relevance of the previously discerned physical dependence (and then we have a series of formulations 1-2-3) or as modifications of the very dependence for the same conditions of relevance; in this case we obtain the series 1-2'-3''. In the first case we would state the evolution of knowledge has led to more precise formulations of Ohm's law, that is to more precise definitions of its range of relevance. In the second instance, it led to the replacement of Ohm's law by other ones which were more precise.

By analogy, we could analyze the transition from Boyle's to Van der Waals' law, and either state that Van der Waals formulated a more precise physical relation among the mass, volume and pressure of gases, or we could state that he made more precise the relationship discovered by Boyle, limiting the range of its relevance to ideal gases, that is to gases in which there is no interaction between particles. We could also state that Einstein formulated a new relation between the mass of a body and its velocity, or that he reformulated more precisely an already established law, limiting its range of relevance to movements which are slow with respect to the velocity of light.

Thus, the transition to a more precise formulation, could be presented formally either as the transition from one law to another, or as a process of making the range of relevance of the older law increasingly precise. In this context it makes little difference which presentation we might choose. Before we go on, however, to the question of why we sometimes choose to speak of a new formulation of a law, and at other times of a more precise formulation of an existing one, let us take a closer look at the relationships occurring between the formulations of Ohm's law cited above.

It is not difficult to notice that if in the equation $RI + L(dI/dt) + Q/C = E$, we make $Q/C = 0$, in other words if we assume that the capacity is infinite, we will arrive at the following equation: $RI + L(dI/dt) = E$. If in the last equation we make $L(dI/dt) = 0$, in other words if we assume that there is no self-induction in the given circuit, then in turn we will arrive at Ohm's law: $RI = V_2 - V_1$ (E is the difference in voltage).

We may then state that between the successive formulations (1), (2) and (3), that is between different laws, there occurs such a relation that each successive formulation contains the previous formulation within it. Or we may state the physical dependence expressed in (1) is a particular case of that expressed in (2), which in turn is a particular case of (3). The same relationship occurs, of course, between the formulations (1), (2') and (3"). This, however, concerns not the formulations of the physical dependence, but the descriptions of the range of relevance. And this, in effect, is the relationship which we previously termed a formal correspondence. It occurs regardless of whether the empirical meaning of the terms used in the preceding formulations change or remain the same. There exists a formal correspondence between Ohm's law and the differential equation $RI + L(dI/dt) = E$, irrespective of whether the concepts of voltage, current and resistance do or do not maintain the same empirical interpretation within both models. The same relationship occurs between the laws of Van der Waals and Boyle, the theories of Newton and Einstein, and the classical and the quantum theories of radiation.

If we agree that the birth of relativistic mechanics marks a revolution in science, but the successive reformulations of Ohm's law do not, then we must by the same token agree that the occurrence of a formal correspondence does not permit us to distinguish between revolutionary and non-revolutionary changes in science. It is not true that formal correspondence is only a feature of evolutionary change, that is to say, that scientific revolutions are always characterized by a break in this relationship. In any case, in the argument as to whether a scientific revolution does or does not mark a break in the correspondence between theories, it is certain that it is not formal correspondence which is at issue.

IV

Let us assume that we have accepted a law $\prod_x W(x) \to Z(x)$, and that it is discovered that an object belonging to the set x does not fulfill the dependence $Z(x)$, or, in other words, that an experiment has revealed an anomaly to this law. The first question which arises in such a situation is why it is so, and whether we are able to point to the causes of the fact that the previously confirmed dependence was not fulfilled in this case.

The simplest possibility would be that we made an error in assigning the given object to the set x, and in expecting that it would fulfill the given dependence. Thus, for example, when Davy found that potassium hydroxide decomposes under electrolysis, he faced the alternative: he could either reject the accepted opinion that chemical elements are not decomposable, or assume, against the reigning conviction, that potassium hydroxide is not an element.

Another cause of error may result from the fact that the set x is in reality composed of two sub-sets, x_1 and x_2, and that the elements of x_1 fulfill the dependence $Z(x)$, while those of x_2 do not. As we have had a chance to observe in analysing the formulations of Ohm's law, changes in the content of our knowledge may be expressed in such cases either by reformulating the dependence $Z(x)$ into $Z'(x)$ for the same range of relevance, or by modifying the range of relevance of $Z(x)$, that is by means of reformulating $W(x)$ into $W'(x)$. The extreme case of such an instance would occur when the sub-set x_1 is an empty set, that is when an experiment demonstrates that the dependency $Z(x)$ is not precisely fulfilled in any situation.

It is quite apparent that the answer to the question of why the dependence $Z(x)$ is not fulfilled in the hitherto assumed range of reference, depends on theoretical views concerning the given domain of phenomena under investigation. Thus, for example, it depends on whether we are aware of the phenomenon of self-induction at the time when we state that the Ohm's law is not strictly fulfilled by all conductors.

This, in fact, is a matter of primary importance. It indicates that in formulating the range of relevance of any scientific statement we are never able to indicate all the conditions, both positive and negative, which must be fulfilled by the objects which this statement covers. The delimitation of these objects is dependent on the contemporary theoretical knowledge, *and it cannot transgress this knowledge.* As long as the phenomenon of self-induction remains unknown, it is impossible to formulate the range of relevance of Ohm's law by taking conductors with no self-induction into account. And correspondingly, as long as we do not allow the possibility that the mass of a body may depend on its velocity, there is no way that we can formulate the laws of mechanics in such a manner as to accordingly restrict the range of their relevance. For formulating these restrictions would be tantamount to assuming a possibility of which we are not even aware. In other words, the situation presents itself in such a manner that not only can we never be sure that the actual range of relevance is equal to the range formulated in the scientific law, but, moreover, we are never *able to encompass, in formulating a law, all those conditions, which we actually take into account.* Certain of these conditions may become evident only *ex post*, that is only after the law in its previous formulation has been falsified and replaced by a new one. No law of science, therefore, can adequately formulate the conditions in which the dependency $Z(x)$ is strictly fulfilled.[13] What is more, the description of conditions $W(x)$ can never enumerate all the assumptions which are actually accepted and on which the formulation of its alleged range of relevance depends. A large number of these assumptions are accepted quite unconsciously and lie dormant in the background knowledge. Therefore, at the moment when we discover an empirical anomaly, we have to question not only all the assumptions which have been explicitly expressed in the formulation of the law [in $W(x)$] and in the theory which explains it, but also those which we have accepted implicitly, without fully realizing that we have actually taken them into account. The character of the change in our knowledge which results from the elimination of the anomaly, depends on which of our assumptions we are forced to modify.

Just as Faraday's investigations on induction and self-induction allowed for the more precise formulation of Ohm's law by limiting its range of relevance to conductors with no self-induction, so the realization that the mass of a body is directly dependent on the ratio of its velocity to the speed of light, permitted the more precise formulation of the range of relevance of the laws of non-relativistic mechanics. In both cases the more precise formulation of laws was made possible through modifications in theoretical knowledge. In both cases, the old and the new theory were linked by the relationship of formal correspondence; and, finally, in both cases, the new formulation could explain all those natural phenomena which were explained by the old, as well as those anomalies with which the preceding theory could not come to grips. And yet, for reasons already mentioned, (that is because the discoveries of J. Henry, M. Faraday and F. Lenz did not change the empirical meaning of the basic assumptions of the theory of electricity, while the theory of relativity did cause a reinterpretation of the basic categories of mechanics), in the relativistic case we are dealing with a revolution in the content of knowledge while in the case of Ohm's law, with evolutionary change. The linguistic custom according to which we sometimes speak of a more precise formulation of the law, while at other times of the discovery of a new relation of dependence, seems to be at least partially connected with the aforementioned difference in the character of the modification in our knowledge.

Therefore, as I see it, we should single out three types of modifications in the content of our knowledge:

The first type concerns cases when the refutation and replacement of the empirical statement[14] does not result in a semantic reinterpretation of the concepts of the theory which serves as the *explicans* of both statements. The transformations in the formulation of Ohm's law, which we have analyzed above, may serve as an example of this type of transformation.

The second type concerns cases when the elimination of an empirical anomaly requires a semantic reinterpretation of the concepts of the theory which hitherto served to explain the observed phenomena. It concerns cases when a new *explanans* is incommensurable in content with the older one, with the restriction, however, that the reinterpretation concerns only those assumptions which are limited in their application to certain, more or less broadly defined special fields of investigation, and not to science as a whole.

And finally, the third type concerns situations such that the elimination of empirical anomalies requires the reinterpretation of concepts which are so general that this reinterpretation affects the whole field of scientific investigation, and in time changes the total world-view.

NOTES

[1] R. Suszko, 'Formal Logic and the Development of Knowledge', in *Problems in The Philosophy of Science*, (ed. I. Lakatos and A. Musgrave) Amsterdam 1968, pp. 210–222.

[2] *Ibid.*, p. 215.
[3] *Ibid.*, p. 217.
[4] *Ibid.*, p. 218.
[5] *Ibid.*, p. 219.
[6] *Ibid.*, p. 220.
[7] *Ibid.*, p. 220.
[8] See J. Giedymin's remarks to Suszko's paper, *Ibid.*, pp. 225–227.
[9] R. Suszko, *ibid.*, p. 230.
[10]. See A. Koyré, *Études galiléennes*, Paris 1966, especially pp. 83–161.
[11] See W. Mejbaum, 'Prawo i sformułowania' (Scientific Law and its Formulations), in S. Amsterdamski, Z. Augustynek, and W. Mejbaum, eds., *Prawo, konieczność, prawdopodobieństwo* (*Laws, Necessity and Probability*), Warsaw 1964, pp. 227-253. Cf. also J. Such, *O uniwersalności praw nauki* (*On the universal character of scientific laws*), Warsaw 1972, Chapter 4.
[12] W. Mejbaum, *Ibid.*, p. 236.
[13] The range of relevance may turn out to be inadequate in two ways; it may either be formulated too narrowly or too widely. Experiments pertaining to the alleged range of relevance may demonstrate its inadequacy only in the second case. In the first case, the inadequacy is discovered as a rule *via the theory*, i.e. by means of a reinterpretation of theoretical concepts which changes their denotations. After this reinterpretation, they denote phenomena which they had not previously denoted. Such a change of the range of relevance (its widening) may result in a situation such that some of the empirical data which neither confirmed nor falsified the previous formulation of the law will be now regarded as anomalies with respect to the new formulation.

CHAPTER VIII

REVOLUTIONS IN SCIENCE:
SCIENCE AND PHILOSOPHY

I

When we take a position contrary to the stance of radical empiricism, and accept the view that every scientific statement is theoretical in character, we cannot, then, accept the thesis that semantic correspondence remains preserved through all changes in the content of knowledge. Certain authors who accept this conclusion speak of the incommensurability of successive theories and treat such changes in the content of knowledge as the specific feature of scientific revolutions. Thus, for example, according to Kuhn, pre- and post-revolutionary paradigms provide just such non-corresponding, incommensurable world-perspectives. But if there is not, and cannot be, any supra-paradigmatic instance which could serve as a 'neutral' frame of reference for the appraisal of competing paradigms, then the revolutionary transition from the old to the new point of view cannot be a fully rational process. (The old paradigm no longer provides any criteria of rationality, while the new one is as yet incapable of providing them.) In order to explain revolutionary change, we must reach out beyond methodology and refer to the psychology of research.

Criticizing this point of view, Popper writes:

I do admit that at any moment we are prisoners caught in the framework of our theories; our expectations; our past experiences; our language. But we are prisoners in a Pickwickian sense: if we try, we can break out of our framework at any time. Admittedly, we shall find ourselves again in a framework, but it will be a better and a roomier one; and we can at any moment break out of it again.

The central point is that a critical discussion and a comparison of the various frameworks is always possible. It is just a dogma – a dangerous dogma – that the different frameworks are like mutually untranslatable languages .... The Myth of the Framework is, in our time, the central bulwark of irrationalism.[1]

Let us now note that if scientific theories were indeed closed and coherent languages, and if the introduction of every new concept led to a new world-perspective, then each and every change in the content of knowledge should be treated as a revolution since it would break the semantic correspondence between theories and make them mutually untranslatable, incompatible with one another.

We are thus faced with the question: having accepted the fact that at least some of the changes in the content of our knowledge are revolutions, are we also compelled to accept the position that the transition from the old to the new point of view, which is incommensurable with its predecessor, occurs irrationally? Or, in other words, is the thesis on the incommensurability of successive points of view equivalent with the assertion that there is no *rational* path leading from the one to the other, *that there are no broader frameworks embracing both of the competing, incommensurable theories*?

In order to accept Popper's critical observation, cited above, we could have either to demonstrate that there exists some superior point of view in respect to which we may evaluate non-corresponding theories, or at least to demonstrate that there is some broader consensus within which the transition may occur rationally. However, as we have had occasion to demonstrate in previous chapters, neither Popper's logic of scientific discovery, nor Lakatos' methodology of research programs are capable of adequately solving this problem. We have seen that the methodological rules cannot constitute that superior point of view, first, because they themselves may be endangered by a crisis in the event of a scientific revolution, and, second, for the reason that the criterion of demarcation isolates science from the broader frameworks within which the transition from one point of view to the next occurs. If, following Popper, we reject the radical empiricist solution to the problem of an empirical basis of science and undertake the problem of the mechanism of the evolution of knowledge, we cannot stop half-way and continue to treat science on the basis of the criterion of demarcation, as an isolated fragment of our convictions. Once the doors are opened, they cannot remain half-open, nor be shut again. The guard who stands at these doors, armed with a suprahistorical criterion of demarcation must be retired from his post. Scientific theories are neither closed and coherent languages, so that every change in our knowledge constitutes a revolution, nor are they based on observational sentences with unchangeable meaning (dependent exclusively on experiment) which would guarantee the correspondence in content between successive theories.

I do not believe, however, that the non-existence of a supra-historical

point of view in science, which would always allow us to rationally appraise non-corresponding theories, would be paramount to the thesis that
the transition from one theory to another is an irrational process. Both
Popper and Kuhn incorrectly identify these two problems with each other.
Accepting Kuhn's position that pre- and post-revolutionary theories do
not correspond to each other in content, a position which is a direct consequence of Popper's criticism of the belief in a purely empirical basis of
science, we are by no means forced to conclude that there exists no rational
manner of transition from one theory to another. On the other hand, in
accepting Popper's position that this transition is in fact a rational process, we are not obliged to accept the opinion that this is guaranteed by a
supra-historical point of view provided by the logic of scientific discovery.
The fact that criteria of rationality are historical in character does not
imply that they do not exist at all. The real question is, what are these
frameworks within which scientific revolutions take place, and how do
they condition the rational transition from the old point of view to the
new one?

II

In criticizing Kuhn's position in Chapter VI, we drew attention to the
ambiguity of the term 'scientific revolution' in his conception. According
to what we have said, it seems imperative to differentiate between those
changes which affect the whole knowledge of a given period, and those
which concern only particular fragments of that knowledge, i.e. between
'global' and 'local' revolutions. Disturbances in one area of the field, to
use Quine's metaphor once again, spread and radiate to neighboring areas,
but do not constitute a disturbance of the whole field. If we look at the
field as a whole, there is always a 'revolution' in some area, and in this
sense science is always in a state of crisis. Global revolutions, on the other
hand rarely occur in science. It seems that the differentiation between these
two phenomena, which unfortunately are often united by one term, may
introduce some clarity into the problem of the mechanism of transition
from one 'point of view' to another. Many misunderstandings which have
gathered around this question are most obviously linked to the ambiguity
of the term 'scientific revolution'. Thus, it is obvious that anyone who
understands a scientific revolution to be what Suszko has termed a 'weak
revolution', must see the factors conditioning such phenomena in a differ-

ent light from someone who uses the term 'revolution' to denote the transition from one research program to another within the realm of a particular discipline, or from someone (as for example A. Koyré[2]) who speaks of a global revolution in the whole of our knowledge.

There can be no doubt that Koyré is referring to global changes in our knowledge, and in the style of scientific thinking – a style which is determined, first of all, according to his opinion, by reigning philosophical ideas. In other words, he refers to such changes as occurred in the 16th and 17th centuries and which, quite possibly, we have been witnessing ever since the last decades of the 19th century, but not to such phenomena as the introduction into a theory of a new concept or, for example, the substitution of one theory by another within the frameworks of the same style of scientific thinking. It goes without saying that global revolutions are realized through local ones, that they are impossible without them, and that only by the study of the latter are we able to discover them. However, the opposite does not hold true; changes in one area of the field are not necessarily caused by changes in the whole nor do they always affect the whole of the field. For the scope of the revolution depends upon convictions which were involved in the practice of a given discipline and in the practice of science as a whole, and which now are undermined by the revolution.

The distinction between these two different types of revolution is closely linked with what we have said above (Chapter II, Section VI) about the function which science fulfills in human culture. If this function is not fully taken into account, then not only are the differences between these two types of phenomena blurred and, as in Kuhn's case, the types contaminated with each other, but it also becomes impossible to study their mutual relationships. This, in turn, must lead to an excessive, even absolute stress upon the element of discontinuity in the evaluation of science, to the treatment of each local revolution as if it constituted a total break of *all* (and not only of some) of the convictions common to all scientists in a given epoch. What then may go unnoticed is that the more general paradigmatic convictions which pertain not only to the narrow field which is undergoing a crisis, may constitute a relatively stable consensus, which may help rationally to overcome the crisis. In distinguishing between revolutions on the basis of their scope, it is not enough to state that a revolution consists of a change of the paradigm and that the pre- and

post-revolutionary points of view are incommensurable, do not correspond to one another. It is also important to attempt to discern which
paradigmatic convictions are totally questioned or rejected, and which ones
remain, in spite of the crisis. For these convictions, or points of view which
are shared by all the investigators in a given field and which are not affected at the moment of crisis, may constitute the frameworks within which
the crisis may be rationally overcome, within which communication is
possible. The more fundamental (that is the more involved in the practice
not only of one narrow discipline, but of larger fields of knowledge, or of
science in general) are the convictions questioned by the crisis, the broader
the scope of the revolution, and the more global its character.

We have already noted that there are no such convictions which cannot
be questioned in a crisis, whether they be methodological rules or criteria,
principles of rational research procedure, the concept of experience or
that of truth, etc. Therefore, when we look into the past, and especially
into the remote past, we cannot see how – without falsifying history – we
can describe the transition, which after all did take place, especially when
the frameworks of contemporary convictions, which enabled this transition to take place, have themselves eroded away. In order to understand
how the transition took place, how it was possible that men, who were
known widely as the leading minds of their times, first held views which
appear to us to be 'myths', only to change them for other, to us no less
'mythological' convictions, it is necessary to reconstruct not a supra-
historical logic of evolution, but the opinions (both 'scientific' and 'non-
scientific') commonly held at the time, when the particular fragment of
knowledge under study underwent a crisis. We must search out which
rational roads of agreement and transition to a new point of view were
open at the time, and how scholars proceeded along these paths so that
they managed to overcome the crises and become 'enslaved' by the new
paradigm.

In each case of crisis, there existed a higher court of appeal. And if we
are unable to discern it, it is only for the reason that the accepted criterion
of demarcation excluded it from the range of factors affecting the evolution of scientific knowledge. At the time of crisis within a specific discipline, scholars do not cease in their discussions; indeed, quite to the
contrary, their arguments grow in intensity. And there is no reason to
believe that they cannot communicate with and understand one another.

But this superior authority is itself a part of the field, and not something external to it. Therefore, it is temporary and may itself be threatened by the crisis. It is an arbitrator that may continue to judge, until the time when it itself is now subject to judgment.

It is not enough to state that paradigms contain certain 'non-scientific' beliefs – philosophical or for that matter metaphysical – which condition a particular manner of understanding the world, and the human experience. For it is necessary to solve the problem of the role of these beliefs both in the practice of normal science and in the overcoming of crises which it encounters. If one distinguishes between types of revolutions according to their scope and according to the character of beliefs whose modifications they imply, if one takes in account not only the negative role of 'non-scientific' convictions (even if they are unavoidable), but also their regulatory role as relatively stable frameworks within which more or less global crises in science are overcome, then, by that token, we may avoid the alternative: either to understand science as a continuous process determined by unchangeable criteria of rationality, or to treat revolutions as irrational leaps from one enslavement to another?

III

What then constitutes those broader frameworks within which crises in science, resulting either from internal contradictions between theories or from conflict with experiments, may be rationally overcome? We have already said that the break of semantic correspondence between successive scientific theories may constitute the criterion for distinguishing between evolutionary and revolutionary changes. It is much more difficult, however, to differentiate between what I have termed 'local' and 'global' scientific revolutions.

We have noted above that it is possible to distinguish a local from a global revolution by taking into consideration the nature of the concepts whose change the revolution implies. In the case of a local revolution, these would be concepts whose scope of application is limited to certain, more or less broadly defined, fields of inquiry, while in the case of global revolution, concepts which are involved in science as a whole, that is in the study of any discipline. This characterization, however, is definitely too general and of little help if it does not indicate at the same time – and not simply by enumeration – which concepts and related beliefs are of concern here, and what role they fulfill within science.

The first, but not the sufficient, step in the direction of distinguishing this class of concepts which are involved in the practice of science in general, could be the indication that in every period of the history of science there exists a particular discipline which functions as a *basic discipline*. In the 18th and 19th centuries, mechanics was considered to be such a discipline, while today it is physics which fulfills this role in the structure of our knowledge. (In order to avoid any misunderstandings, I would like to stress that the thesis concerning the existence of a basic discipline within the structure of knowledge may, but does not have to, constitute the basis of reductionism, i.e. of the point of view that postulates the reduction of all scientific statements to statements of the basic discipline. We shall not, however, enter into this problem here.[3])

The statement that physics is the *basic discipline* means, first of all, that every really existing object is a physical object and, therefore, possesses certain qualities which are studied by physics. This means that physics studies, among others, certain objects which have no qualities studied by any other empirical sciences. In the course of its development, physics becomes more and more capable of treating every object or process studied by other empirical sciences as particular *physical structures*; this means that it is able to treat them as entities which as wholes may be characterised by specific, particular qualities studied by a given specialized discipline, as well as structures composed of elements which do not have these qualities. In this sense, every object, each material structure, regardless of its particular qualities which make it an object of inquiry of some scientific discipline, has a physical aspect, i.e, *it must be describable in physical terms*. In this respect a neutron, an atom, a chemical particle of water, a molecule of protein, an organic cell, as well as the human and the animal brain, a living organism and society are physical structures. This does not mean, or imply of course, that physical descriptions may fully exhaust their analysis and provide us with the full knowledge of these objects.

There are two methodological postulates which follow from the statement that physics is a basic science, that every material structure is a physical structure, and that every natural process is a physical process.

First, every science, in studying even the most particular qualities of material structures, qualities which are its specialized object of investigation, must also take into account the basic qualities and interactions of material objects established by physics. To present a rather commonplace example, in studying any biological process, the biologist must take into account a number of physical laws such as, for instance, the law of the conservation of energy.

Second, each science which studies even the most particular processes which take place within specific material structures, must search for the physical basis of these processes.

Assuming, then, the existence of a basic discipline, we could state that global revolutions are, above all, connected with changes of those concepts which are indispensable for the characterization of every object of scientific inquiry, in other words of those concepts which codetermine the ontology which lies at the base of the scientific enterprise. And it is obvious that not only the content of the concepts, but also their very ensemble change in the process of the evolution of knowledge.

From this point of view, global revolutions are always linked with radical modifications of the world perspectives provided by the basic discipline, and bring with them local revolutions in particular fields of research. The scientific revolution of the 16th and 17th centuries was prompted, as is well known, by radical changes in the cosmological vision of the universe which was then based on the physics and philosophy of Aristotle. In turn it triggered off changes in particular scientific disciplines such as mechanics, optics, chemistry, etc. The total result of this revolution was the formulation of a new ontology which had its source in Newtonian mechanics. The breakdown of this ontology under the successive blows of Maxwell's electromagnetic theory, Planck's quantum theory, and Einstein's theory of relativity led finally, at the turn of this century, to the next global revolution, one which we are still witnessing, and which, in all likelihood, has not yet run its course. The decades of work which Einstein dedicated to the construction of the unified field theory was a titanic effort to climax this revolution with a coherent theory which would account for the most fundamental ontological features of the material world. Whatever are the sources of the modifications in ontology, they in turn cause disturbances which radiate steadily onto the whole field of our knowledge. In order to grasp this phenomenon, it is only necessary to reflect on the consequences which were drawn by biology, chemistry or even psychology from the mechanistic theory of nature in the 16th and

17th centuries, or for that matter which were taken over from quantum mechanics by contemporary biology and successively by other disciplines of the science of man.

Thus, we may state that one of the basic features distinguishing local from global revolutions, consists in the fact that while the former lead to the reorganization of a limited field of inquiry (as, for example, the revolution in chemistry in the 17th century), not effecting changes in the ontological vision of the universe as a whole, and quite to the contrary are often a result of such changes, the latter are characterized by that very fact that they do cause radical changes in this vision.

We must stress, however, once more, that the differentiation between a basic discipline and its conceptual categories, by means of which we define the ontological structure of the universe of human experience can only be the first, not the final step leading towards the specification of a set of concepts and beliefs involved in the practice of science in general.

IV

Besides the concepts which are necessary for the ontological characterization of the structure of the universe as an object of cognition, a similar role in this respect is played by conceptions of man as a knowing subject. Perhaps in a less overt and immediate manner, but no less important than the ontological visions of the universe, they co-determine the common frameworks within which the cognitive effort is undertaken. Moreover, they serve to determine the rules of scientific procedures, and the changes which affect these concepts just as the similar changes in the ontological outlook have important consequences for the whole field of scientific investigation. Together with ontological theses, they constitute what might be referred to as the style of scientific thinking, the heuristics, or the regulatory principles of science.[4] For global revolutions in science are characterized not only by the introduction of a new order into the sphere of human experience, but also by changes in the cognitive outlook of man upon the universe and upon himself: in other words, both a new vision of the universe and a new conception of the cognitive experience.

If we search, then, for those common conceptual frames within which it is possible to overcome local crises in science, and whose steady erosion leads in turn to global revolutions, then besides changes in the ontological vision of the universe, we must take into account changes in knowledge of man, and first of all, in the knowledge of the functioning of the mind. Global revolutions are, at the same time, revolutions in the philosophy of man: this was indeed the case during the revolution of the 16th and 17th centuries, and such is indubitably the case today.

Characterizing by one, shortened formula the intellectual revolution of the 16th and 17th centuries, it is possible to state that God, as the measure of all things, was replaced by Man. Man, however, as a knowing subject, was provided with at least some divine attributes. He was to be an ideal observer external to the universe under study, and he was to be capable of achieving the absolute truth about this universe. Beginning with Descartes and Bacon, and ending with Kant and Hegel, this concept of man, variously justified in philosophical terms as being capable of cognitive procedures, co-determined the style of thinking of modern science. This may be discerned within both empiricist as well as rationalist epistemology. And it was this conception, as I have already said, which articulated the concept of sense-experience, of truth and falsehood, of the object, and of the possible limits of human cognition.

By including the knowing subject more and more into the world of nature, by depriving it of its privileged, outside status within nature, modern science steadily undermined its own epistemological basis. Now not God, and not Man standing beyond nature and confronting it as a perfect knowing subject, but nature itself was to become the measure of all things. The theory of relativity, and quite paradoxically if we take into account its name, states the absolute relevance of the laws of nature which are to be formulated in such a manner as to be true and knowable for every knowing subject, with the restriction that the subject is not transcendent with respect to nature.[5]

Therefore, if we look at changes in the style of scientific thinking, or at the regulatory principles of actual science from the indicated point of view, then we must agree that it is not only the revolutions in the basic discipline and in its fundamental conceptual apparatus determining the ontological world perspective that take part in this process, but also changes in a much broader area of scientific knowledge – in logic, mathematics, neurophysiology, linguistics, or sociology. If scientific cognition consists in the occurrence of some relationship between the object and the knowing subject, then cognition is impossible without some assumptions concerning both the subject and the object. These assumptions may be ignored only as long as the cognitive effort does not reach the level of self-reflection. This reflection, however, becomes indispensable when the principles, which until that moment have guided our procedure, no longer lead to success. It is perhaps superfluous to say that, as opposed to Kant, we do not consider these principles self-evident. Rejecting those views which attempt to find the absolute, unquestionable foundation of cognition either in the sphere of experience or in the human mind, we are forced to treat the regulatory principles of actual science as the product of the human cognitive effort and accept the bewildering fact that on the grounds of achieved results, the human mind is capable of undertaking to criticize the very principles due to which it has reached these results.

V

It has been customary to term that set of assumptions and concepts which determines the global vision of the universe and man's cognitive stance towards it as metaphysics. In a certain sense these convictions are indeed untestable. They cannot be confirmed or rejected on the basis of an individual experiment as scientific statements are to be. However, the untestability of metaphysical assumptions in the course of scientific procedure necessitates two remarks.

First of all, I would say that it is one thing to state that all metaphysical beliefs are untestable and quite another to claim that all untestable sentences are metaphysical. The utterance 'This book is the work of the devil' is, admittedly, untestable. It would, however, be an exaggeration to label this sentence as metaphysical. Similarly, just as we would not label every testable sentence (such as 'Joe has freckles') as scientific, so not every untestable sentence deserves to be termed as metaphysical. Both these qualifications also depend on the role which these statements play in the system of knowledge. Even if we tried to demarcate one field from the other, the testability would be, at best, a necessary but not a sufficient condition for the demarcation of these two domains of beliefs. For we see some difference between the role that may be played in the evolution of knowledge by the untestable beliefs quoted above and the role which was played, for instance, by the Aristotelian doctrine of form and matter or the Cartesian theory of *cogito*.

Second, if we do not treat the testability of metaphysical assumptions, in the framework within which science is practiced, as a problem which may be solved by an individual experiment, then the problem of their testability, at least within science, becomes more complicated. This kind of untestability does not imply that "it is impossible in general to speak rationally about a certain kind of corroboration, or rather about the decreasing or increasing degree of corroboration of ontological hypotheses by the evolution of scientific knowledge in a given historical period." Nor does it imply that "it is impossible to claim that our cognitive attitude towards ontological hypothesis cannot rationally depend upon the results of experiments."[6] We believe *mutatis mutandis* that this opinion concerns not only ontological hypotheses but also epistemological and methodological convictions. For, as the history of science teaches us, it is undoubtedly the case that hitherto accepted metaphysical assumptions must at some point be rejected. This is not so because a single experiment has demonstrated their falsehood, for this is simply impossible, but for the reason that they are no longer capable of fulfilling their heuristic function *within science*. They do not allow for the unification of science into one rational whole, nor provide the indications of how to respond to local revolutions which are caused in certain parts of the field by conflicts with experiment, nor allow for the critical analysis of the foundations of science. No one ever experimentally refuted either the Aristotelian doctrine of form and matter, or the Cartesian theory of *cogito*, Locke's conception of the *tabula rasa* or Kant's idea of *a priori* forms and categories of cognition. And nevertheless, each of these conceptions was in a certain sense 'falsified' within science, rejected as it ceased to fulfill its heuristic function *in relation to science*, and ceased to be useful.

It is possible to state, following Kant,[7] that metaphysics is a battlefield on which the human mind battles for the justification of its own assumptions with which it approaches research. But it is also possible to reverse the metaphor and argue that science is a battlefield on which metaphysical systems which provide the empirically untestable justification for particular manners of research procedure and which create the frameworks for their critical analysis, have to demonstrate their usefulness as regulatory principles for the construction of internally consistent systems of knowledge which conform to experimental data.

If, as I suppose, untestable metaphysical convictions do, in fact, play the role of regulatory principles in the sense that has been outlined above, then they cannot be tested by this or that experiment, but rather by the 'epistemological experience',[8] that is by the actual course of the evolution of science – by its successes and failures. It is on this battlefield that global heuristic piograms, which postulate the requirements that scientific theories have to fulfill, compete with each other in a given period and prove themselves to be productive or not. They may, for instance, require the construction either of deterministic or statistical theories; of phenomeno-logical, that is those limited to the description of experimental data, or of hypothetical theories which presuppose the actual existence of objects and interactions which are not given in direct observation. They may also require the construction of theories which guarantee the objective (that is, independent of the relationship between subject and object) description of reality, or give up this requirement on the grounds of some epistemo-logical assumptions.

It is necessary to note that it is not true that the global heuristic pro-grams can function in science only if the regulatory principles on which they are based are true, that is if the ontological and epistemological hypothesis which they comprise, are true. It is, therefore, not the case that the regulatory principle requiring the construction of deterministic the-ories may be fruitfully applied in science only if the ontological thesis of strict determinism is true. As we recognize today that the thesis of strict determinism is false, we cannot deny that the utilization of the regulatory principle which was based on it was, and in certain disciplines still remains, heuristically fruitful. A sufficient condition of its heuristic fruitfulness *within certain fields of research* may be, for example, the truth of a 'weaker' ontological hypothesis according to which the observed regularity of phenomena are based on sharp statistical distributions, or the truth of the even weaker presupposition which only assumes the oc-currence of regularities. There can be little doubt that no concrete experi-ment is capable of solving which of these two competing ontological hypo-theses is true. But is it equally untrue that, given the fact that neither of these hypotheses is testable, there can be no rational way for their apprais-al, that our attitude toward them cannot depend on the results of pre-vious epistemological experience. On this basis, that is on the basis of successes and failures resulting from research directed by these regulatory principles, our trust in them may grow stronger oi weaker, although we are not capable of demonstrating their truth; and, moreover, we must treat them as untestable in the sense in which scientific statements are to be testable.

It is equally impossible to propose any methodological rules (see our criticism of Lakatos, presented in Chapter VI) which would serve to indicate after what period of failures the regulatory principles, or even the whole heuristic program, is to be discarded. It is nevertheless a fact that although a scientific experiment cannot falsify the regulatory principles and the epistemological experience does not compel us to reject them, nevertheless, inappropriate metaphysical assumptions are, in due course, rejected and replaced by others.

The persecutors of metaphysics in science notice only that the accepted metaphysical assumptions are groundless. They fail to see, however, that while they are untestable, the very same mechanism by which they are introduced into science serves to eliminate them at the moment that they cannot fulfill their heuristic function. The program of elimination of metaphysical theses from science would be justified (which does not yet mean that it would be possible to realise it) if it were, in fact, the case that their acceptance or rejection would in each case be inevitably arbitrary, that is if there existed within science no mechanism for their rational evaluation, choice and elimination. I am convinced that this is not the case. First of all, because without the acceptance of some heuristic program which provides historically changeable criteria of rationality and methodological rules of research procedure, the practice of science is simply impossible. Second, it is not so because metaphysics in science is not simply a set of unfalsifiable theses but a complex of regulatory principles which make up the heuristic programs and which, although on the basis of different rules and on a different 'level' of experience, are subject to rational evaluation in respect to their fruitfulness. For it is not true that rational criticism in science is limited to criticism by experiment and observation.

The second remark about the global heuristic programs with which metaphysics provides science concerns the fact that these programs postulate only some of the qualities which are to characterize scientific theories concerning particular domains of phenomena. In other words, no such program univocally designates a theory which would fulfill all its requirements within the given field of inquiry (and as we have noted in Chapter

VI, this fact is not taken into account by Kuhn in his conception of paradigms and normal science). Therefore, it is possible to build various, non-contradictory, mutually competing research programs within the given discipline on the grounds of the same global heuristic program. By the same token, the concept of global heuristic programs by no means implies such a vision of normal science as is to be found in Kuhn's work. It is possible to reconcile it with the existence of several smaller or larger schools within various disciplines, with the competition between these schools, as well as with the rational overcoming of local crises in particular fields of inquiry (rational in respect to the global heuristic program).

And finally, a last remark on this question. The concept of the relationship of science to metaphysics (or more simply – philosophy) which I have sketched here is founded on the conviction that philosophical theses (ontological and epistemological) foim the basis for the formulation of normative principles regulating research procedures. These principles make up the heuristic programs, which lead to success and failure, and which are judged on this basis to be eventually retained or rejected. This opinion requires some further explanation, for otherwise, against the intentions of the author, it might be interpreted as an expression of the instrumentalist treatment of philosophy. The simplest, and at the same time sufficient manner of dissipating any doubts in the matter would be to indicate that we have spoken here of the *relationship* of philosophy to science, of the role which philosophical claims play in science, and, therefore, the very manner of posing the question already implies such an answer. Nowhere, however, was it stated that the *only* manner of evaluating philosophical claims is by means of the *function* which they play in intellectual life, for instance, in science. Such a conclusion could only be drawn by someone who equates the problem of meaningfulness with that of testability. For if philosophical statements are to fulfill these functions within science, then it is for the reason that they are meaningful statements, that they express something about the universe. If this were not so, then they could not serve as a foundation for the formulation of heuristic programs.

VI

The opinion presented above concerning the relationship of science and philosophy may create the impression that it is based on a *petitio principii* argument. The regulatory principles deriving from philosophy are considered at one point as the product of the evolution of cognition, and at another as its indispensable condition. We had already noted this circumstance earlier, when we spoke of the fact that the human mind is capable of taking under criticism the principles which have led it to certain results. If knowledge without presuppositions is impossible – since in our opinion it is not provided either by experience or by the mind – then how is the evolution of knowledge, founded on the systematic criticism of its own assumptions, possible?

I can see no other answer to this question then the one which has been signaled many times on the pages of this book – that our knowledge *never* constitutes a closed and coherent system. It would be a closed system if, in the first place, all the assumptions actually accepted at a given moment were accepted *explicitly* and, second, if the criticism of these assumptions itself was possible without any assumptions. It would constitute a coherent system if all the (implicitly and explicitly) accepted assumptions were noncontradictory. In the light of Gödel's theorem, we know that even a mathematical system of knowledge cannot escape incompleteness and internal contradiction. First of all, it is possible to formulate within any deductive system sentences which cannot be deduced from its axioms. The introduction of new axioms may provide the solution for some hitherto unsolved problems, but as a result new, unsolved problems will arise. Second, it is impossible to prove the noncontradictory character of such systems exclusively on the grounds of the rules of inference of these systems.[9]

If, therefore, on the basis of a certain system of assumptions it is possible to arrive at a statement whose truth cannot be demonstrated within the system, i.e. without new assumptions, then the phenomenon of the change of assumptions ceases to be so bewildering. To the contrary, it turns out to provide the condition for the solution of problems which were begotten by the accepted assumptions. And, it is not the case that our conviction (that the accepted assumptions are both at the same time the product and the condition for our knowledge) results in a vicious circle; and neither is it the case that if this conviction were to be true then the accepted assumptions could not undergo any change and that the system would itself turn in a 'circle'.

176 176176 176176176176176 176176176176176176176176176176176176176176176176176176176176176

BIBLIOGRAPHY

d'Abro, A., *The Evolution of Scientific Thought from Newton to Einstein*, New York 1927.

Agassi, J., *Towards an Historiography of Science*, The Hague 1963.

Agassi, J., 'The Nature of Scientific Problems and their Roots in Metaphysics', in *The Critical Approach to Science and Philosophy*, ed. by M. Bunge, Glencoe 1964, pp. 189–211.

Agassi, J., 'Science in Flux', in *Boston Studies in the Philosophy of Science*, Vol. III, ed. by R. S. Cohen and M. W. Wartofsky, Dordrecht 1967, pp. 293–323.

Ajdukiewicz, K., 'Das Weltbild und die Begriffsapparatur', *Erkenntnis* 4 (1934), 259–287.

Ajdukiewicz, K., 'W sprawie artykułu prof. A. Schaffa o moich poglądach filozoficznych' (On Prof. A. Schaff's Article on my Philosophical Opinions), in *Język i poznanie* [*Language and Cognition*], Vol. II, Warsaw 1965, pp. 155–191.

Ajdukiewicz, K., *Logika Pragmatyczna* [*Pragmatic Logic*], Warsaw 1965.

Amsterdamski, S., 'O obiektywnych interpretacjach pojęcia prawdopodobieństwa' (On the Objective Interpretations of the Concept of Probability), in *Prawo, konieczność, prawdopodobieństwo* [*Laws, Necessity and Probability*], Warsaw 1964, pp. 1–125.

Amsterdamski, S., 'Prawdziwość i prawdopodobieństwo' (Truth and Probability), *Zeszyty Naukowe Uniwersytetu Łódzkiego*, Warsaw 1967, pp. 1–15.

Amsterdamski, S., 'Historia nauki i filozofia nauki' (The History of Science and the Philosophy of Science), postscript to the Polish edition of T. S. Kuhn, *The Structure of Scientific Revolutions*, Warsaw 1968, pp. 189–206.

Amsterdamski, S., 'Scjentyzm i rewolucja naukowo-techniczna' (Scientism and the Scientific-technical Revolution), *Zagadnienia Naukoznawstwa* 23 (1970), 16–34.

Amsterdamski, S., 'Spór o problem postępu w historii nauki' (The Debate over Progress in the History of Science), *Kwartalnik Historii Nauki i Techniki*, 1970, pp. 487–506.

Amsterdamski, S., 'Nauka i wartości' (Science and Values), *Zagadnienia Naukoznawstwa* 25 (1971), 58–73.

Amsterdamski, S., 'Science as a Object of Philosophical Reflection', *Organon* 9 (1973), 35–60.

Ayer, A. J., *Language, Truth and Logic*, London 1946.

Ayer, A. J., *The Problem of Knowledge*, London 1956.

Bachelard, G., *La formation de l'esprit scientifique*, Paris 1959.

Barber, B., 'Resistance by Scientists to Scientific Discovery' in *The Sociology of Science*, ed. by B. Barber and W. Hirsch, Glencoe 1962, pp. 539–556.

Barber, B., 'Tension and Accommodation between Science and Humanism', *American Behavioral Scientist* 7 (1963), 3–8.

Bartley, W. W., 'Theories of Demarcation between Science and Metaphysics', in *Problems in the Philosophy of Science*, Vol. III, Amsterdam 1968, pp. 40–64.

Bartley, W. W., 'Reply', *ibid.*, pp. 102–119.

Bernal, J. D., *The Social Function of Science*, London 1939.

Bernard, C., *An Introduction to the Study of Experimental Medicine*, New York 1957.

Bochner, S., *The Role of Mathematics in the Rise of Science*, Princeton 1966.

Bohr, N., 'Discussions with A. Einstein on Epistemological Problems in Atomic Physics', in *Albert Einstein – Philosopher Scientist*, ed. by P. A. Schilpp, New York 1959, pp. 199–242.

Brunet, P., *L'introduction des théories de Newton en France au XVIII siècle*, Vol. I, Paris 1931.

Butterfield, H., *The Origins of Modern Science, 1500–1800*, New York 1957.

Carnap, R., 'Überwindung der Metaphysik durch logische Analyse der Sprache', *Erkenntnis* 2 (1931), 219–241.

Carnap, R., 'Testability and Meaning', *Philosophy of Science* 3 (1936), 419–471; 4 (1937), 1–40.

Carnap R., *Logical Foundations of Probability*, Chicago 1951.

Carnap, R., 'The Methodological Character of Theoretical Concepts' in *Minnesota Studies in the Philosophy of Science*, Vol. 1, ed. by H. Feigl and M. Scriven, Minneapolis 1956, pp. 38–75.

Carnap, R., *Philosophical Foundations of Physics*, New York 1966.

Conant, J., *Modern Science and Modern Man*, New York 1952.

Copernicus, N., *The Complete Works*, Vol. 1, London 1972.

Duhem, P., *The Aim and Structure of Physical Theory*, New York 1962.

Eilstein, H., 'Hipotezy ontologiczne i orientacje ontologiczne' (Ontological Hypotheses and Ontological Orientations), in *Teoria i Doświadczenie* [*Theory and Experience*] ed. by H. Eilstein and M. Przełęcki, Warsaw 1966, pp. 223–243.

Einstein, A. and Infeld, L., *The Evolution of Physics*, New York 1938.

Einstein, A., 'Considerations Concerning the Fundamentals of Theoretical Physics', *Science* 91 (1940), 487–492.

Einstein, A., 'Reply to Criticism', in *Albert Einstein – Philosopher – Scientist*, ed. by P. A. Schilpp, New York 1959, pp. 665–688.

Einstein, A., 'Autobiographical Notes', *ibid.*, pp. 2–95.

Enriques, F., 'Signification de l'histoire de la pensée scientifique', *Actualités scientifiques et industrielles* 161 (1934).

Feigl, H., 'Operationism and Scientific Method', *Psychological Review* 52 (1945), 250–259.

Feigl, H., 'Some Major Issues and Developments in the Philosophy of Science of Logical Empiricism', in *Minnesota Studies in the Philosophy of Science*, Vol. I ed. by H. Feigl and M. Scriven, Minneapolis 1956, pp. 3–35.

Feigl, H., 'Philosophy of Science' in *Philosophy*, ed. by R. M. Chisholm, H. Feigl, W. K. Frankena, J. Passmore, and M. Thompson, Englewood Cliffs 1964, pp. 465–540.

Feyerabend, P. K., 'Explanation, Prediction and Empiricism', in *Minnesota Studies in the Philosophy of Science*, Vol. III, ed. by H. Feigl and G. Maxwell, Minneapolis 1962, pp. 28–97.

Feyerabend, P. K., 'Problems of Empiricism', in *Beyond the Edge of Certainty*, ed. by R. Colodny, Englewood Cliffs, 1965, pp. 145–260.

Feyerabend, P. K. 'Reply to Criticism', in *Boston Studies in the Philosophy of Science*, Vol. II, ed. by R. S. Cohen and M. Wartofsky, Dordrecht 1965, pp. 223–261.

Feyerabend, P. K., 'Consolation for a Specialist', in *Criticism and the Growth of Knowledge*, ed. by I. Lakatos and A. Musgrave, Cambridge 1970, pp. 197–230.

Geymonat, L., *Filozofia i filozofia nauki* (Polish trans. from the Italian), Warsaw 1966.

Giedymin, J., 'O teoretycznym sensie tzw. zdań obserwacyjnych' (On the Theoretical Sense of So-called Observational Statements), in *Teoria i doświadczenie* [*Theory and Experience*], ed. by H. Eilstein and M. Przełęcki, Warsaw 1966, pp. 91–110.

Giedymin, J., 'Indukcjonizm i antyindukcjonizm' (Inductionism and Anti-inductionism), in *Logiczna teoria nauki* [*The Logical Theory of Science*], ed. by T. Pawłowski, Warsaw 1966, pp. 269–318.

Giedymin, J., 'Revolutionary Changes, Non-translatability and Crucial Experiments', in *Problems in the Philosophy of Science*, Vol. III, ed. by I. Lakatos and A. Musgrave, Amsterdam 1968, pp. 223–226.

Grünbaum, A., 'Geometry, Chronometry and Empiricism', in *Minnesota Studies in the Philosophy of Science*, Vol. III, ed. by H. Feigl and G. Maxwell, Minneapolis 1962, pp. 407–527.

Grünbaum, A., 'The Falsifiability of a Component of a Theoretical System', in *Mind, Matter and Method: Essays in Philosophy of Honor of H. Feigl*, ed. by P. K. Feyerabend and G. Maxwell, Minneapolis 1966, pp. 273–305.

Grünbaum, A., 'Can We Ascertain the Falsity of a Scientific Hypothesis?' in *Observation and Theory in Science*, Baltimore 1971.

Grünbaum, A., 'Falsifiability and Rationality', manuscript, 1971.

Habermas, J., *Knowledge and Human Interests*, Boston 1971.

Hall, A. R., *Scientific Revolution 1500–1800: The Formation of the Modern Scientific Attitude*, 2nd edn. London 1962.

Hall, H. H., 'Scientists and Politicians', in *The Sociology of Science*, ed. by B. Barber and W. Hirsch, Glencoe 1962, pp. 269–287.

Heisenberg, W., *Physics and Philosophy*, New York 1958.

Hempel, C. G., 'Problems and Changes in the Empiricist Criterion of Meaning', *Revue internationale de Philosophy* 4 (1950), 41–63.

Hempel, C. G., 'The Theoretician's Dilemma', in *Minnesota Studies in the Philosophy of Science*, Vol. II, ed. by H. Feigl, Minneapolis 1958.

Hesse, M., 'Reasons and Evaluation in the History of Science', in *Changing Perspectives in the History of Science: Essays in Honor of Joseph Needham*, ed. by M. Teich and R. Young, London and Boston 1973, pp. 127–147.

Holton, G. J., ed., *Science and Culture: A Study of Cohesive and Disjunctive Forces*, Boston 1965.

Holton, G. J., *Thematic Origins of Scientific Thought: Kepler to Einstein*, Cambridge 1973.

Ingarden, R., 'Główne tendencje neopozytywizmu' (The Main Tendencies of Neopositivism), in *Z Badań nad filozofią współczesną* [*Studies in Contemporary Philosophy*], Warsaw 1963, pp. 643–654.

Ingarden, R., 'Próba przebudowy filozofii przez neopozytywistów' (The Neopositivist Attempt at Changing Philosophy), *ibid.*, pp. 655–662.

Jacob, F., *The Logic of Life*, New York, 1974.

Jammer, M., *Concepts of Space*, Cambridge 1957.

Jammer, M., *Concepts of Mass*, Cambridge 1961.

Jammer, M., *The Conceptual Development of Quantum Mechanics*, Cambridge 1966.

Kant, I., *Critique of Pure Reason*, Garden City 1966.

Kołakowski, L., *Notatki o współczesnej kontrreformacji* [*Notes on Contemporary Counterreformation*], Warsaw 1962.

Kołakowski, L., *Kultura i fetysze* [*Culture and Fetish*], Warsaw 1967.

Kołakowski, L., *The Alienation of Reason: A History of Positivist Thought*, New York 1968.

Kołakowski, L., *Die Gegenwärtigkeit des Mythos*, Munich 1973.

Kotarbińska, J., 'Kontrowersja: dedukcjonizm – indukcjonizm' (The Controversy:

Deductionism – Inductionism), in *Logiczna teoria nauki* [*The Logical Theory of Science*], ed. by T. Pawłowski, Warsaw 1966, pp. 319–341.

Kotarbińska, J., 'Ewolucja koła wiedenskiego' (The Development of the Vienna Circle), *ibid.*, pp. 295–318.

Koyré, A., *From the Closed World to the Infinite Universe*, Baltimore 1957.

Koyré, A., *Études d'histoire de la pensée philosophique*, Paris 1961.

Koyré, A., *Newtonian Studies*, London 1965.

Koyré, A., *Études Galiléennes*, Paris 1966.

Kuhn, T. S., *The Copernican Revolution*, Cambridge 1957.

Kuhn, T. S., 'Measurements in Modern Physics', *Isis* **52** (1961), 161–193.

Kuhn, T. S., *The Structure of Scientific Revolutions*, 2nd edn, enlarged, Chicago 1970.

Kuhn, T. S., 'Logic of Discovery or Psychology of Research', in *Criticism and the Growth of Knowledge*, ed. by I. Lakatos and A. Musgrave, Cambridge 1970, pp. 2–27.

Kuhn, T. S., 'Reflections on my Critics', *ibid.*, pp. 231–278.

Kuhn, T. S., 'Mathematical *versus* Experimental Traditions in the Development of Science', George Sarton Memorial Lecture, Washington, D.C., December 28, 1972, manuscript.

Kuznetsov, B. G., *Princip otnositel'nosti v antičnoj, klassičeskoj i kvantowoj fizike* [*The Principle of Relativity in Ancient, Classical and Quantum Physics*], Moscow 1959.

Kuznetsov, B. G., *Einstein*, Moscow 1963.

Kuznetsov, B. G., *Galileo*, Moscow 1964.

Lakatos, I., 'Proofs and Refutations', *British Journal for the Philosophy of Science* **14** (1963), 1–25; 120–139; 221–243; 296–342.

Lakatos, I., 'Changes in the Problem of Inductive Logic', in *The Problem of Inductive Logic*, ed. by I. Lakatos, Amsterdam 1968, pp. 315–417.

Lakatos, I., 'Falsification and the Methodology of Scientific Research Programmes', in *Criticism and the Growth of Knowledge*, ed. by I. Lakatos and A. Musgrave, Cambridge 1970, pp. 91–195.

Lakatos, I., 'History and its Rational Reconstructions', in *Boston Studies in the Philosophy of Science*, Vol. VIII, ed. by R. C. Buck and R. S. Cohen, Dordrecht 1972, pp. 91–138.

Lakatos, I., 'Replies to Critics', *ibid.*, pp. 174–182.

Lakatos, I. and Zahar, E., 'Why did Copernicus' Program Supercede Ptolemy's?' manuscript.

Lanczos, C., *Albert Einstein and the Cosmic World Order*, New York 1965.

Lanczos, C., 'Rationalism and the Physical World', in *Boston Studies in the Philosophy of Science*, Vol. III, ed. by R. S. Cohen and M. Wartofsky, Dordrecht 1967, pp. 181–198.

Leibniz, G. W., *The Leibniz-Clarke Correspondence*, ed. by H. G. Alexander, Manchester 1956.

Le Roy, E., 'Science et philosophie', *Revue de métaphysique et de morale* **7** (1899), 375–425; 503–562; 706–731.

Le Roy, E., 'Un positivisme nouveau', *Revue de métaphysique et de morale* **9** (1901), 138–153.

Mach, E., *The Science of Mechanics*, Chicago 1960.

Madden, E., *The Structure of Scientific Thought*, London 1968.

Mannheim, K., *Ideology and Utopia*, London 1936.

Masterman, M., 'The Nature of the Paradigm' in *Criticism and the Growth of Knowledge*, ed. by I. Lakatos and A. Musgrave, Cambridge 1970, pp. 59–90.

Mehlberg, H., 'O niesprawdzalnych założeniach nauki' (On the Unverifiable Assumptions of Science), in *Logiczna teoria nauki* [The Logical Theory of Science], ed. by T. Pawłowski, Warsaw 1966, pp. 341–362.

Mehlberg, H., *The Reach of Science*, Toronto 1959.

Mejbaum, W., 'Prawo i sformułowania' (Laws and Formulations), in *Prawo, konieczność, prawdopodobieństwo* [*Laws, Necessity and Probability*], ed. by S. Amsterdamski, Z. Augustynek, and W. Mejbaum, Warsaw 1964, pp. 225–256.

Merton, R. K., 'Priorities in Scientific Discovery: A Chapter in the Sociology of Science', in *The Sociology of Science*, ed. by B. Barber and W. Hirsch, Glencoe 1962, pp. 447–485.

Mullins, N. C., *A Sociological Theory of Normal and Revolutionary Science*, manuscript.

Musgrave, A., 'On a Demarcation Dispute', in *Problems in the Philosophy of Science*, Vol. III, ed. by I. Lakatos and A. Musgrave, Amsterdam 1968, pp. 67–78.

Nagel, E. and Newman, J. R., *Gödel's Proof*, New York 1958.

Nagel, E., *The Structure of Science*, New York 1961.

Neugebauer, O., *The Exact Sciences in Antiquity*, Boston 1957.

Nowak, L., 'Metodologia nauki w świetle historii nauk przyrodniczych' (The Methodology of Science in the Light of the History of the Natural Sciences), *Studia Filozoficzne*, 1970, no. 4–5, pp. 296–301.

Poincaré, H., *Science and Hypothesis*, New York 1952.

Poincaré, H., *Science and Method*, New York 1952.

Poincaré, H., *The Value of Science*, New York 1958.

Polanyi, M., *Personal Knowledge*, London 1958.

Polanyi, M., *The Tacit Dimension*, Garden City 1967.

Pomian, K., 'Le cartésianisme, les Erudits et l'histoire', *Archiwum Historii Filozofii i Myśli Społecznej* 12 (1966), 175–204.

Pomian, K., 'Działanie i sumienie' (Action and Conscience), *Studia Filozoficzne*, 1967, No. 3., pp. 21–68.

Pomian, K., *Przeszłość jako przedmiot wiary* [*The Past as an Object of Faith*], Warsaw 1968.

Popper, K. R., *The Poverty of Historicism*, London 1957.

Popper, K. R., *The Logic of Scientific Discovery*, London 1959.

Popper, K. R., 'Remarks on the Problem of Demarcation and of Rationality', in *Problems in the Philosophy of Science*, Vol. III, ed. by I. Lakatos and A. Musgrave, Amsterdam 1968, pp. 88–101.

Popper, K. R., 'Epistemology without a Knowing Subject', in *Proceedings of the IIIrd International Congress of Logic, Methodology and Philosophy of Science*, Amsterdam 1968, pp. 333–373.

Popper, K. R., *Objective Knowledge*, Oxford 1972.

Przełęcki, M., 'O definiowaniu terminów spostrzeżeniowych' (On Defining Observational Terms), in *Rozprawy logiczne. Księga pamiątkowa ku czci K. Ajdukiewicza* [*Studies in Logic: Essays in Honor of K. Ajdukiewicz*], Warsaw 1964, pp. 155–184.

Przełęcki, M., 'W sprawie istnienia przedmiotów teoretycznych' (On the Subject of the Existence of Theoretical Objects), in *Teoria i doświadczenie* [*Theory and Experience*], ed. by H. Eilstein and M. Przełęcki, Warsaw 1966, pp. 49–67.

Quine, W. V. O., *Word and Object*, Cambridge 1960.

Quine, W. V. O., 'Two Dogmas of Empiricism', in *From a Logical Point of View*, Cambridge 1961, pp. 20–46.

Quine, W. V. O., 'Carnap and Logical Truth', in *Logic and Language: Studies Presented* to Professor Rudolf Carnap on the Occasion of His Seventieth Birthday, Dordrecht 1962, pp. 39–63.

Rainko, S., 'Epistemologia Diachroniczna' (Diachronic Epistemology), Studia Filozoficzne, 1967, No. 1., pp. 3–40.

Rainko, S., Rola podmiotu w poznaniu [The Role of the Subject in Cognition], Warsaw 1972.

Ravetz, J. R., 'Tragedy in the History of Science', in Changing Perspectives in the History of Science: Essays in Honor of Joseph Needham, ed. by M. Teich and R. Young, London and Boston 1973, pp. 204–222.

Reichenbach, H., Experience and Prediction, Chicago 1938.

Reichenbach, H., The Rise of Scientific Philosophy, Berkeley 1951.

Reichenbach, H., The Direction of Time, Los Angeles 1956.

Scheffler, I., The Anatomy of Inquiry, New York 1963.

Schlick, M., 'Meaning and Verification', in Readings in Philosophical Analysis, ed. by H. Feigl and M. Sellars, New York 1949, pp. 146–174.

Schrodinger, E., Science et humanisme, Paris 1954.

Singer, C., A Short History of Scientific Ideas, Oxford 1959.

Snow, C. P., The Two Cultures, Cambridge 1961.

Snow, C. P., Two Cultures, A Second Look, New York 1961.

da Solla Price, D. J., Science Since Babylon, New Haven 1961.

Stanosz, B., 'Introduction' to the Polish edition of Quine, W. V. O., From a Logical Point of View, Warsaw 1969.

Such, J., O universalnosci praw nauki, Warsaw 1972.

Suszko, R., 'Logika formalma i niektóre zagadnienia teorii poznania' (Formal Logic and Certain Problems of Epistemiology), in Logiczna teoria nauki [The Logical Theory of Science], ed. by T. Pawłowski, Warsaw 1966, pp. 505–578.

Suszko, R., 'Formal Logic and the Development of Knowledge', in Problems in the Philosophy of Science, Vol. III, ed. by I. Lakatos and A. Musgrave, Amsterdam 1968, pp. 218–222; 227–230.

Teske, A., 'Wolterowskie Elementy fizyki Newtona', Introduction to Voltaire, Elementy fizyki Newtona, Warsaw 1956.

Teske, A., The History of Physics and the Philosophy of Science: Selected Essays, Warsaw 1972.

Toulmin, S. E., 'Conceptual Revolutions in Science', in Boston Studies in the Philosophy of Science, Vol. III, ed. by R. S. Cohen and M. Wartofsky, Dordrecht 1967, pp. 311–347.

Toulmin, S. E., Human Understanding, Princeton 1972.

Toulmin, S. E., 'Does the Distinction between Normal Science and Revolutionary Science Hold Water?', in Criticism and the Growth of Knowledge, ed. by I. Lakatos and A. Musgrave, Cambridge 1970, pp. 39–48.

Vavilov, S., Isaak Newton, Moscow 1945.

Vermel, E. M., Istorija učenija o kletje [The History of the Study of the Cell], Moscow 1970.

Voltaire, F. M., Eléments de la philosophie de Newton,, mis à la portée de tout le monde, Amsterdam 1738.

Wartofsky, M., 'Metaphysics as Heuristic for Science', in Boston Studies in the Philosophy of Science, Vol. III, ed. by R. S. Cohen and M. Wartofsky, Dordrecht 1967, pp. 132–172.

Watkins, J. W. N., 'Against Normal Science', in Criticism and the Growth of Knowledge, Vol. V, ed. by I. Lakatos and A. Musgrave, Cambridge 1970, pp. 25–38.

INDEX OF NAMES

184   INDEX OF NAMES

Heraclides 14
Hempel, C. G. 45, 179
Hesse, M. 179
Henıy, J. 158
Hirsch, W. 142, 177, 179, 181
Holton, G. J. 179
Hoyle, F. 44
Hume, D. 25, 57, 69–72
Husserl, E. 51

Infeld, L. IX, 178
Ingarden, R. IX, 30, 46, 179

Jacob, F. 13, 22–3, 179
Jammer, M. 179

Kandinsky, W. 5
Kant, I. VII, 20, 25, 26, 50, 72, 84–5,
     169–71, 176, 179
Kepler, J. 15, 66, 134, 179
Klee, P. 5
Klein, M. XVI
Kołakowski, L. IX, 45–6, 142, 179
Kotarbińska, J. 78, 86–7, 179–80
Kotarbiński, T. IX
Koyré, A. 17–8, 21–3, 151, 159, 176, 180
Kramers, H. A. 130
Krzywicki, L. IX
Kuhn, T. S. X, XI, 22–3, 38, 46, 53–5, 62,
     67, 87, 95, 113, 117–29, 135–6, 139–43,
     147, 160–3, 173, 177, 180
Kuznjetzov, B. G. 22, 87, 180

Lakatos, I. X, XVI, 33, 46, 62–8, 95,
     100–3, 113, 117, 129–43, 147, 158, 173,
     176–82
Lanczos, C. 180
Lange, O. IX
Laplace, P. S. 14, 18, 44
Laudan, L. XVI
Leibniz, G. W. 16–20, 23, 180
Le Roy, E. 72, 180
Lenz, F. 158
Leonardo da Vinci 5, 152
Leśniewski, S. IX
Leverrier, U. 97
Lobatschevski, N. 114
Locke, J. 71–2, 171
Łukasiewicz, J. IX

Mach, E. 25, 69, 71–2, 180
Madden, E. 180
Malebranche, N. 10, 22
Mannheim, K. 46, 180
Masterman, M. 119, 141, 180
Maxwell, J. C. 14, 44, 71, 98, 167
Maxwell, G. 178–9
Mehlberg, H. 27, 45, 180–1
Mejbaum, W. X, 152, 154, 159, 181
Mendel, G. 5, 23
Mendeleer, D. I. 22
Merton, R. K. 181
Michałowski, P. XVI
Mill, J. S. 74
Morgan, T. H. 44
Mostowski, A. IX
Mullins, N. C. 181
Musgrave, A. 113, 143, 158, 176, 178–82

Nagel, E. XV, 46, 176, 181
Napoleon 18
Needham, J. 179, 182
Neugebauer, O. 22
Newman, J. R. 176
Newton, I. 5, 11–2, 14, 16–20, 43–4,
     97–8, 109, 123, 134–5, 155, 177–8
Nowak, L. 181

Ohm, G. 153–8
Oppenheimer, R. 127
Osiander, A. 15, 23

Passmore, J. 178
Pavlov, I. P. 44
Pawłowski, T. 86, 179–82
Planck, M. 148, 167
Plato 51, 55, 118
Poincaré, H. 25, 72, 88, 102, 107–10, 112,
     114, 134, 181
Polanyi, M. 55, 67, 95, 129, 141, 181
Pomian, K. X, XVIII, 46, 87, 181
Popper, K. R. X, 25, 27, 31–3, 45–59,
     67–8, 72, 76–9, 84–96, 100–3, 111–4,
     117, 121, 125, 127, 128, 132, 139, 140–3,
     160–2, 176, 181
Przełęcki, M. 87, 178, 181
Ptolemy 5, 8, 14, 15

Quine, W. V. O. 35, 45–6, 108, 114, 162,
     182

INDEX OF SUBJECTS

SYNTHESE LIBRARY

Monographs on Epistemology, Logic, Methodology,
Philosophy of Science, Sociology of Science and of Knowledge, and on the
Mathematical Methods of Social and Behavioral Sciences

Managing Editor:

JAAKKO HINTIKKA (Academy of Finland and Stanford University)

Editors:

ROBERT S. COHEN (Boston University)
DONALD DAVIDSON (The Rockefeller University and Princeton University)
GABRIËL NUCHELMANS (University of Leyden)
WESLEY C. SALMON (University of Arizona)

1. J. M. BOCHEŃSKI, *A Precis of Mathematical Logic.* 1959, X + 100 pp.
2. P. L. GUIRAUD, *Problèmes et méthodes de la statistique linguistique.* 1960, VI + 146 pp.
3. HANS FREUDENTHAL (ed.), *The Concept and the Role of the Model in Mathematics and Natural and Social Sciences, Proceedings of a Colloquium held at Utrecht, The Netherlands, January 1960.* 1961, VI + 194 pp.
4. EVERT W. BETH, *Formal Methods. An Introduction to Symbolic Logic and the Study of Effective Operations in Arithmetic and Logic.* 1962, XIV + 170 pp.
5. B. H. KAZEMIER and D. VUYSJE (eds.), *Logic and Language. Studies Dedicated to Professor Rudolf Carnap on the Occasion of his Seventieth Birthday.* 1962, VI + 256 pp.
6. MARX W. WARTOFSKY (ed.), *Proceedings of the Boston Colloquium for the Philosophy of Science, 1961–1962,* Boston Studies in the Philosophy of Science (ed. by Robert S. Cohen and Marx W. Wartofsky), Volume I. 1973, VIII + 212 pp.
7. A. A. ZINOV'EV, *Philosophical Problems of Many-Valued Logic.* 1963, XIV + 155 pp.
8. GEORGES GURVITCH, *The Spectrum of Social Time.* 1964, XXVI + 152 pp.
9. PAUL LORENZEN, *Formal Logic.* 1965, VIII + 123 pp.
10. ROBERT S. COHEN and MARX W. WARTOFSKY (eds.), *In Honor of Philipp Frank,* Boston Studies in the Philosophy of Science (ed. by Robert S. Cohen and Marx W. Wartofsky), Volume II. 1965, XXXIV + 475 pp.
11. EVERT W. BETH, *Mathematical Thought. An Introduction to the Philosophy of Mathematics.* 1965, XII + 208 pp.
12. EVERT W. BETH and JEAN PIAGET, *Mathematical Epistemology and Psychology.* 1966, XII + 326 pp.
13. GUIDO KÜNG, *Ontology and the Logistic Analysis of Language. An Enquiry into the Contemporary Views on Universals.* 1967, XI + 210 pp.
14. ROBERT S. COHEN and MARX W. WARTOFSKY (eds.), *Proceedings of the Boston Colloquium for the Philosophy of Science 1964–1966, in Memory of Norwood Russell Hanson,* Boston Studies in the Philosophy of Science (ed. by Robert S. Cohen and Marx W. Wartofsky), Volume III. 1967, XLIX + 489 pp.

15. C. D. Broad, *Induction, Probability, and Causation. Selected Papers.* 1968, XI + 296 pp.
16. Günther Patzig, *Aristotle's Theory of the Syllogism. A Logical-Philosophical Study of Book A of the Prior Analytics.* 1968, XVII + 215 pp.
17. Nicholas Rescher, *Topics in Philosophical Logic.* 1968, XIV + 347 pp.
18. Robert S. Cohen and Marx W. Wartofsky (eds.), *Proceedings of the Boston Colloquium for the Philosophy of Science 1966–1968,* Boston Studies in the Philosophy of Science (ed. by Robert S. Cohen and Marx W. Wartofsky), Volume IV. 1969, VIII + 537 pp.
19. Robert S. Cohen and Marx W. Wartofsky (eds.), *Proceedings of the Boston Colloquium for the Philosophy of Science 1966–1968,* Boston Studies in the Philosophy of Science (ed. by Robert S. Cohen and Marx W. Wartofsky), Volume V. 1969, VIII + 482 pp.
20. J. W. Davis, D. J. Hockney, and W. K. Wilson (eds.), *Philosophical Logic.* 1969, VIII + 277 pp.
21. D. Davidson and J. Hintikka (eds.), *Words and Objections: Essays on the Work of W. V. Quine.* 1969, VIII + 366 pp.
22. Patrick Suppes, *Studies in the Methodology and Foundations of Science. Selected Papers from 1911 to 1969,* XII + 473 pp.
23. Jaakko Hintikka, *Models for Modalities. Selected Essays.* 1969, IX + 220 pp.
24. Nicholas Rescher et al. (eds.). *Essay in Honor of Carl G. Hempel. A Tribute on the Occasion of his Sixty-Fifth Birthday.* 1969, VII + 272 pp.
25. P. V. Tavanec (ed.), *Problems of the Logic of Scientific Knowledge.* 1969, XII + 429 pp.
26. Marshall Swain (ed.), *Induction, Acceptance, and Rational Belief.* 1970. VII + 232 pp.
27. Robert S. Cohen and Raymond J. Seeger (eds.), *Ernst Mach; Physicist and Philosopher,* Boston Studies in the Philosophy of Science (ed. by Robert S. Cohen and Marx W. Wartofsky), Volume VI. 1970, VIII + 295 pp.
28. Jaakko Hintikka and Patrick Suppes, *Information and Inference.* 1970, X + 366 pp.
29. Karel Lambert, *Philosophical Problems in Logic. Some Recent Developments.* 1970, VII + 176 pp.
30. Rolf A. Eberle, *Nominalistic Systems.* 1970, IX + 217 pp.
31. Paul Weingartner and Gerhard Zecha (eds.), *Induction, Physics, and Ethics, Proceedings and Discussions of the 1968 Salzburg Colloquium in the Philosophy of Science.* 1970, X + 382 pp.
32. Evert W. Beth, *Aspects of Modern Logic.* 1970, XI + 176 pp.
33. Risto Hilpinen (ed.), *Deontic Logic: Introductory and Systematic Readings.* 1971, VII + 182 pp.
34. Jean-Louis Krivine, *Introduction to Axiomatic Set Theory.* 1971, VII + 98 pp.
35. Joseph D. Sneed, *The Logical Stricture of Mathematical Physics.* 1971, XV + 311 pp.
36. Carl R. Kordig, *The Justification of Scientific Change.* 1971, XIV + 119 pp.
37. Milič Čapek, *Bergson and Modern Physics,* Boston Studies in the Philosophy of Science (ed. by Robert S. Cohen and Marx W. Wartofsky), Volume VII, 1971, XV + 414 pp.
38. Norwood Russell Hanson, *What I do not Believe, and other Essays,* ed. by Stephen Toulmin and Harry Woolf, 1971, XII + 390 pp.

39. ROGER C. BUCK and ROBERT S. COHEN (eds.), *PSA 1970. In Memory of Rudolf Carnap*, Boston Studies in the Philosophy of Science (ed. by Robert S. Cohen and Marx W. Wartofsky, Volume VIII. 1971, LXVI + 615 pp. Also available as a paperback.
40. DONALD DAVIDSON and GILBERT HARMAN (eds.), *Semantics of Natural Language*. 1972, X + 769 pp. Also available as a paperback.
41. YEHOSHUA BAR-HILLEL (ed)., *Pragmatics of Natural Languages*. 1971, VII + 231 pp.
42. SÖREN STENLUND, *Combinators, λ-Terms and Proof Theory*. 1972, 184 pp.
43. MARTIN STRAUSS, *Modern Physics and Its Philosophy. Selected Papers in the Logic, History, and Philosophy of Science*. 1972, X + 297 pp.
44. MARIO BUNGE, *Method, Model and Matter*. 1973, VII + 196 pp.
45. MARIO BUNGE, *Philosophy of Physics*. 1973, IX + 248 pp.
46. A. A. ZINOV'EV, *Foundations of the Logical Theory of Scientific Knowledge (Complex Logic)*, Boston Studies in the Philosophy of Science (ed. by Robert S. Cohen and Marx W. Wartofsky), Volume IX. Revised and enlarged English edition with an appendix, by G. A. Smirnov, E. A. Sidorenko, A. M. Fedina, and L. A. Bobrova 1973, XXII + 301 pp. Also available as a paperback.
47. LADISLOV TONDL, *Scientific Procedures*, Boston Studies in the Philosophy of Science (ed. by Robert S. Cohen and Marx W. Wartofsky), Volume X. 1973, XII + 268 pp. Also available as a paperback.
48. NORWOOD RUSSELL HANSON, *Constellations and Conjectures*, ed. by Willard C. Humphreys, Jr. 1973, X + 282 pp.
49. K. J. J. HINTIKKA, J. M. E. MORAVCSIK, and P. SUPPES (eds.), *Approaches to Natural Language. Proceedings of the 1970 Stanford Workshop on Grammar and Semantics*. 1973, VIII + 526 pp. Also available as a paperback.
50. MARIO BUNGE (ed.), *Exact Philosophy – Problems, Tools, and Goals*. 1973, X + 214 pp.
51. RADU J. BOGDAN and ILKKA NIINILUOTO (eds.), *Logic, Language, and Probability. A selection of papers contributed to Sections IV, VI, and XI of the Fourth International Congress for Logic, Methodology, and Philosophy of Science, Bucharest, September 1971*. 1973, X + 323 pp.
52. GLENN PEARCE and PATRICK MAYNARD (eds.), *Conceptual Chance*. 1973, XII + 282 pp.
53. ILKKA NIINILUOTO and RAIMO TUOMELA, *Theoretical Concepts and Hypothetico-Inductive Inference*. 1973, VII + 264 pp.
54. ROLAND FRAÏSSÉ, *Course of Mathematical Logic – Volume I: Relation and Logical Formula*. 1973, XVI + 186 pp. Also available as a paperback.
55. ADOLF GRÜNBAUM, *Philosophical Problems of Space and Time*. Second, enlarged edition, Boston Studies in the Philosophy of Science (ed. by Robert S. Cohen and Marx W. Wartofsky), Volume XII. 1973, XXIII + 884 pp. Also available as a paperback.
56. PATRICK SUPPES (ed.), *Space, Time, and Geometry*. 1973, XI + 424 pp.
57. HANS KELSEN, *Essays in Legal and Moral Philosophy*, selected and introduced by Ota Weinberger. 1973, XXVIII + 300 pp.
58. R. J. SEEGER and ROBERT S. COHEN (eds.), *Philosophical Foundations of Science. Proceedings of an AAAS Program, 1969*. Boston Studies in the Philosophy of Science (ed. by Robert S. Cohen and Marx W. Wartofsky), Volume XI. 1974, X + 545 pp. Also available as paperback.
59. ROBERT S. COHEN and MARX W. WARTOFSKY (eds.), *Logical and Epistemological*

Studies in Contemporary Physics, Boston Studies in the Philosophy of Science (ed. by Robert S. Cohen and Marx W. Wartofsky), Volume XIII. 1973, VIII + 462 pp. Also available as paperback.

60. ROBERT S. COHEN and MARX W. WARTOFSKY (eds.), *Methodological and Historical Essays in the Natural and Social Sciences. Proceedings of the Boston Colloquium for the Philosophy of Science, 1969–1972*, Boston Studies in the Philosophy of Science (ed. by Robert S. Cohen and Marx W. Wartofsky), Volume XIV. 1974, VIII + 405 pp. Also available as paperback.

61. ROBERT S. COHEN, J. J. STACHEL, and MARX W. WARTOFSKY (eds.), *For Dirk Struik. Scientific, Historical and Political Essays in Honor of Dirk J. Struik*, Boston Studies in the Philosophy of Science (ed. by Robert S. Cohen and Marx W. Wartofsky), Volume XV. 1974, XXVII + 652 pp. Also available as paperback.

62. KAZIMIERZ AJDUKIEWICZ, *Pragmatic Logic*, transl. from the Polish by Olgierd Wojtasiewicz. 1974, XV + 460 pp.

63. SÖREN STENLUND (ed.), *Logical Theory and Semantic Analysis. Essays Dedicated to Stig Kanger on His Fiftieth Birthday*. 1974, V + 217 pp.

64. KENNETH F. SCHAFFNER and ROBERT S. COHEN (eds.), *Proceedings of the 1972 Biennial Meeting, Philosophy of Science Association*, Boston Studies in the Philosophy of Science (ed. by Robert S. Cohen and Marx W. Wartofsky), Volume XX. 1974, IX + 444 pp. Also available as paperback.

65. HENRY E. KYBURG, JR., *The Logical Foundations of Statistical Inference*. 1974, IX + 421 pp.

66. MARJORIE GRENE, *The Understanding of Nature: Essays in the Philosophy of Biology*, Boston Studies in the Philosophy of Science (ed. by Robert S. Cohen and Marx W. Wartofsky), Volume XXIII. 1974, XII + 360 pp. Also available as paperback.

67. JAN M. BROEKMAN, *Structuralism: Moscow, Prague, Paris*. 1974, IX + 117 pp.

68. NORMAN GESCHWIND, *Selected Papers on Language and the Brain*, Boston Studies in the Philosophy of Science (ed. by Robert S. Cohen and Marx W. Wartofsky), Volume XVI. 1974, XII + 549 pp. Also available as paperback.

69. ROLAND FRAÏSSÉ. *Course of Mathematical Logic* – Volume II: *Model Theory*. 1974, XIX + 192 pp.

70. ANDRZEJ GRZEGORCZYK, *An Outline of Mathematical Logic. Fundamental Results and Notions Explained with all Details*. 1974, X + 596 pp.

SYNTHESE HISTORICAL LIBRARY

Texts and Studies
in tye History of Logic and Philosophy

Editors:

N. KRETZMANN (Cornell University)
G. NUCHELMANS (University of Leyden)
L. M. DE RIJK (University of Leyden)

1. M. T. BEONIO-BROCCHIERI FUMAGALLI, *The Logic of Abelard.* Translated from the Italian. 1969, IX + 101 pp.

2. GOTTFRIED WILHELM LEIBNITZ, *Philosophical Papers and Letters.* A selection translated and edited, with an introduction, by Leroy E. Loemker. 1969, XII + 736 pp.

3. ERNST MALLY, *Logische Schriften,* ed. by Karl Wolf and Paul Weingartner. 1971, X + 340 pp.

4. LEWIS WHITE BECK (ed.), *Proceedings of the Third International Kant Congress.* 1972, XI + 718 pp.

5. BERNARD BOLZANO, *Theory of Science,* ed. by Jan Berg. 1973, XV + 398 pp.

6. J. M. E. MORAVCSIK (ed.), *Patterns in Plato's Thought. Papers arising out of the 1971 West Coast Greek Philosophy Conference.* 1973, VIII + 212 pp.

7. NABIL SHEHABY, *The Propositional Logic of Avicenna: A Translation from al-Shifā:al-Qiyās,* with Introduction, Commentary and Glossary. 1973, XIII + 296 pp.

8. DESMOND PAUL HENRY, *Commentary on De Grammatico: The Historical-Logical Dimensions of a Dialogue of St. Anselm's.* 1974, IX + 345 pp.

9. JOHN CORCORAN, *Ancient Logic and Its Modern Interpretations.* 1974. X + 208 pp.

10. E. M. BARTH, *The Logic of the Articles in Traditional Philosophy.* 1974, XXVII + 533 pp.

11. JAAKKO HINTIKKA, *Knowledge and the Known. Historical Perspectives in Epistemology.* 1974, XII + 243 pp.

12. E. J. ASHWORTH, *Language and Logic in the Post-Medieval Period.* 1974, XIII + 304 pp.